THE GALACTIC CLUB

UN

AUTRE

MONDE

H^I FOURNIER

PARIS.

THE GALACTIC CLUB

Intelligent Life in Outer Space

Ronald N. Bracewell

STANFORD UNIVERSITY

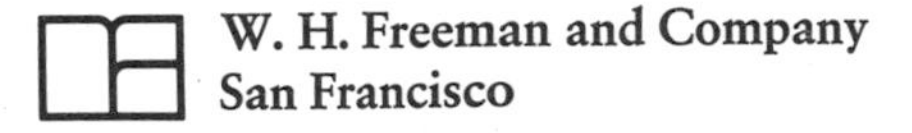

W. H. Freeman and Company
San Francisco

Library of Congress Cataloging in Publication Data

Bracewell, Ronald Newbold, 1921—
The Galactic Club: intelligent life in outer space.

Reprint of the 1974 ed. published by the Stanford Alumni Association, Stanford, Calif., which was issued in series: The Portable Stanford.
Bibliography: p. 130.
1. Life on other planets. I. Titlc.
QB54.B69 1974 919.9′03 74-23056
ISBN 0-7167-0353-X
ISBN 0-7167-0352-1 pbk.

Printed in the United States of America

1 2 3 4 5 6 7 8 9

This book was published
originally as a part of
The Portable Stanford,
a series of books published by
the Stanford Alumni Association.

CONTENTS

PREFACE

The human interest—or anxiety—associated with the possibility of life in outer space is intense. Do we have neighbors? Will we contact them? What will their interest in us be? Friendly? Hostile? Will man expand into space and play a grand future role, bringing fertility to the galaxy? We do not know. This book explores these questions, bringing together the offerings from the different specialties that delve into the possibilities of life in outer space.

College astronomy texts now invariably devote a chapter to what astronomy has to say about life in outer space because the interested layman (and who is not interested?) expects to turn to those who have studied astronomy for reliable comment. Other sciences, especially physics, chemistry, biology, and anthropology, also contribute to the subject of extraterrestrial life; and, in return, the study of these sciences is enriched when it is illustrated by topics such as space travel, development of life, and the establishment of communication between independent civilizations. High school science can only be enlivened by injection of these stimulating and currently developing ideas.

As a participant in this expanding field (see the appended Reader's Guide), much of which has been treated mathematically or described with scientific jargon, I have been well aware of the essential simplicity of the new thoughts as they have come along. Consequently, I am glad that Della Van Heyst persuaded me to write this book; she made it seem easy. Indeed it has proved to be so, thanks to the industry and skills of Cynthia Fry Gunn, whose contributions have far exceeded the standard of ordinary editing.

I am grateful to my daughter Wendy Bracewell for supplying the Russian translations (pp. 101-103) and to my colleagues Pete Allen, Frederick W. Crawford, Werner Graf, John McCarthy, Pierre Noyes, Bernard M. Oliver, and Howard S. Seifert, who kindly read and commented on the manuscript, and to Barry J. Edmonston, Paul R. Ehrlich, Donald Kennedy, Dudley Kirk, Stephen J. Kline, Philip Morrison, Gerald O'Neill, Carl Sagan, Peter A. Sturrock, and Sebastian von Hoerner, who commented on particular chapters.

Stanford, California
October 1974

Ronald N. Bracewell

CREDITS

COVER — Allen, Jesse. *The Ancient River Flows from the Land to Reach the Sea: Midday*. (Central panel of a triptych.) Lithograph. Courtesy of Vorpal Galleries, San Francisco and Chicago.

PAGE ii — Grandville, J.J. *Un Autre Monde*. Paris. 1844.

PAGE xiv — Allen, Jesse. *Panorama with Goats in the Afternoon.* (Left panel of a triptych.) Watercolor. Courtesy of Vorpal Galleries, San Francisco and Chicago.

PAGE 1 — Feynman, Richard P. "The Value of Science." In Edward Hutchings, Jr. (Ed.), *Frontiers of Science: A Survey*. Copyright © 1958 by Basic Books, Inc.

PAGE 4 — M'Guinness, Jim. Other drawings by Jim M'Guinness are found on pages 6, 15, 35, 36, 37, 62, 73, 80, 107, 114, 120, 121, 122, and 124.

PAGE 11 — Allen, Jesse. *Transmigration*. Etching. Courtesy of Vorpal Galleries, San Francisco and Chicago.

PAGE 14 — "Ammonia! Ammonia!" Drawing by R. Grossman. © 1962. The New Yorker Magazine, Inc.

PAGE 16 — Allen, Jesse. *The Ancient River Flows from the Land to Reach the Sea: Midday*. (Right panel.) Lithograph. Courtesy of Vorpal Galleries, San Francisco and Chicago.

PAGE 23 — Allen, Jesse. *Two Macaques: Mid-morning*. Acrylic. Courtesy of Vorpal Galleries, San Francisco and Chicago.

PAGE 25 — Allen, Jesse. *Bird Confronting a Serpent.* Watercolor. Courtesy of Vorpal Galleries, San Francisco and Chicago.

PAGE 26 — *Medieval Fantastic Representation: Man Looking into Outer Space.* Early 16th century woodcut. The Bettman Archive.

PAGE 29 — Escher, M.C. *Three Worlds*. Lithograph. Photograph courtesy of Vorpal Galleries, San Francisco and Chicago.

PAGE 32 — Miró, Joan. *Hair Followed by Two Planets*. Oil on canvas. Courtesy of Pierre Matisse Gallery, New York.

PAGE 40 — *View One*. Artist's concept of ground level view of Cyclops system antennas showing the central control and processing building. Courtesy of NASA/Ames Research Center, Moffett Field, Calif.

View Two. Artist's concept of high aerial view of the entire Cyclops system. Diameter of the antenna array is about 16 kilometers. Courtesy of NASA/Ames Research Center, Moffett Field, Calif.

PAGE 45 — *The Pleiades Cluster, with Nebulosity*. (Crossley reflector.) Courtesy of Lick Observatory, University of California, Santa Cruz.

Jesse Allen, many of whose paintings and lithographs illustrate this book, is a former instructor in French and Italian at Stanford University. Born in Nairobi and educated at Oxford, the self-taught artist now devotes himself entirely to painting.

THE GALACTIC CLUB

CHAPTER ONE

ARE WE ALONE?

Growing in size and complexity
Living things, masses of atoms, DNA, protein
Dancing a pattern ever more intricate.

Out of the cradle onto the dry land
Here it is standing
Atoms with consciousness
Matter with curiosity.

Stands at the sea
Wonders at wondering
I

A universe of atoms
An atom in the universe.

Richard P. Feynman

ONE CAN PICTURE a small clan of our ancestors, geographically isolated for a few generations, wondering whether there were other beings across the water, over the mountains, beyond the desert, or on the other side of whatever barrier limited their range. That is our position today. Our clan is the whole human race; our barrier is space. Is there intelligent life elsewhere in the universe?

Many people think intelligent life does exist somewhere out in space. This is not just an idea of our times, but one which has appeared over and over again throughout man's history. Philosophy, folklore, and art, from all parts of the world throughout the ages, attest to this belief. Some 2,000 years ago the Roman poet Lucretius wrote, "Nature is not unique to the visible world; we must have faith that in other regions of space there exist other earths, inhabited by other people and animals." In the 1500s a good deal of speculation was provoked by Copernicus, who in the year of his death published his view of a sun-centered planetary system. This idea, which demoted the earth from a central and unique position, directly suggested the possibility of other inhabited earths, and the theme was vigorously developed by Giordano Bruno. For this heresy Bruno was burned by the Church in Rome in 1600.

Bruno's views had no factual basis such as modern science requires. Today, the question of the existence of intelligent life has broken through the confines of speculation and has entered the realm of science. Our current theories will be tested in the crucible of expanding scientific data—and therein confirmed or revised. Still the question remains: Is there intelligent life elsewhere in the universe?

What follows in this book is my attempt to illuminate this subject. I will explore and evaluate the ideas of other writers, both scientists and popular non-scientists, germane to our discussion, as well as contribute some ideas of my own.

Other Suns, Other Planets

We begin by asking what evidence there is that planets similar to ours exist. First, since our source of life-giving energy is the sun (which is a star), we need to know whether there are other stars like our sun. We can determine this by measuring the distance to a star and the amount of light and heat received from it. Performing these measurements on many stars, we find that there *are* many other stars like our sun. Assured that sunlight is a commodity in good supply in the starry sky, and knowing that there are over 100 billion stars grouped in our galaxy, the Milky Way, we continue. Are there planets, similar to our earth, circling those stars?

This answer is not as simple. As yet, no planet outside our own solar system has been seen or photographed even with the largest telescopes. If there were planets circling some neighboring star, the tiny points of light shining from them would be overwhelmed by the dazzling glare from the star itself. Prospects are good that planets as large as Jupiter would be detectable by suitable telescopes or spectroscopes mounted on orbiting observatories such as Skylab. But for the present we are forced to pursue indirect lines of argument.

The Origin of Solar Systems

Because of recent astronomical findings, many astronomers expect that most stars, except for the hottest ones, will prove to be accompanied by planets. This expectation is based on a view that stars are born in clouds of interstellar gas and dust that condense under the influence of gravity. A trifling fraction of the gas and dust that escapes being attracted into the forming star remains circling in the planetary orbits. This nebular hypothesis, which in different versions dates back to Kant (1755) and Laplace (1796), was eclipsed in the early 20th century by a theory set forth by Sir James Jeans. He postulated that the planets were formed in a near collision of our sun with another star that was passing by. The chances of a collision of stars in our part of the galaxy are extremely slender—only about one chance in 10 billion per annum. But even the improbable can take place, and if indeed this was the way the earth came into being, to be followed in time by life, the implication was that planets were a rare and accidental occurrence and planetary life even rarer. However, difficulty with the passing star theory arose over the details of how matter would be drawn out of the stars as they passed. According to the theory a near collision brought up huge tidal waves of matter from the passing stars; this matter subsequently condensed into planets. Great tidal deformations can indeed be computed and the breaking away of material predicted, but the subsequent fate of the ejected matter as the stars move apart is to fall back into the stars—not to circle them. This discovery has virtually destroyed the passing star theory. (When you think about this it is not surprising. If you pull something out of a body that is pulling back, the object is not going to come out, turn at right angles, and go into orbit.)

Over a century passed before a serious flaw in the Kant-Laplace nebular hypothesis was removed. Our solar system possesses a puzzling imbalance in the distribution of its *angular momentum.** While 99.9 percent of the mass of the solar system resides in the sun, the angular momentum of the system is mainly in the orbiting planets, 98 percent to be precise. Thus, the sun is turning on its axis at only one fiftieth of the speed it would have if the whole solar nebula had condensed into a single sun unaccompanied by planets. (In an isolated system the angular momentum does not change. This same principle of *conservation of angular momentum* that applies to the whole solar system also explains the striking feat of the pirouetting ice skater who, having built up speed with the arms extended from the shoulders, clasps them to the

* The *angular momentum* of a rotating body is defined as the *angular velocity* multiplied by the *moment of inertia*. (*Angular velocity* is the number of revolutions per minute [RPM]. *Moment of inertia* is obtained by multiplying the mass of each particle of the body by the square of its distance from the axis of rotation and adding the results.)

body and immediately accelerates into a breathtaking spin.)

Such an outcome was not expected under the conditions originally pictured for the contraction of a rotating nebula under its own gravitational attraction. It was thought that, after disconnecting itself from the surrounding interstellar medium which was not to participate in the evolution of the solar system, the irregular nebula would first assume the form of a disk. Then, as the disk contracted to smaller dimensions, which would result in a reduction of the moment of inertia, the angular velocity would of necessity increase, precisely as with the ice skater clasping her elbows. A point would be reached where centrifugal force exceeded gravitational attraction—a rotating ring of matter would then be shed (see *Fig. 1*), leaving a slightly depleted disk to continue shrinking and to shed further concentric rings. It is plausible to suppose that the matter composing these rings would coalesce into single planetary bodies. However, there is nothing in this picture to suggest that such planets would possess any more angular momentum than resided in their parent rings at the time of shedding, which is not enough to account for the total of 98 percent that they are found to have.

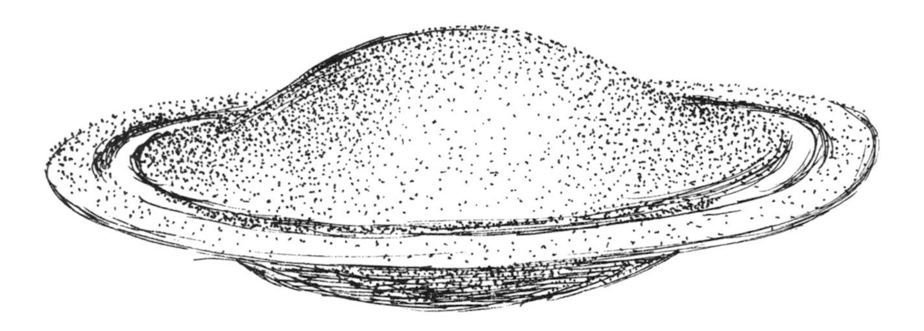

Fig. 1. In this highly schematic drawing, the contracting solar nebula has just shed a ring of matter whose centrifugal force counterbalances gravity. It will remain separate and condense into a planet, while the main mass continues to shrink and shed further rings.

Magnetic Fields

The new development removing this puzzle has been the realization, since the 1950s, that magnetic fields play an important role in astrophysics wherever there is plasma.* I will explain. Considered along with

* Plasma is a gas whose constituent atoms have been smashed into positively and negatively charged components such as protons and electrons.

the solid, liquid, and gaseous states, plasma is the fourth state of matter. Although rare on earth, plasma constitutes most of interplanetary matter and most of our sun. Plasma is in fact the most abundant state of matter observed in the universe as a whole, and is strongly affected by *magnetic fields*. The crucial role played by magnetic fields in the nebular theory is to *feed* angular momentum from the central sun-to-be to the outlying rings. Just as the electric induction motor depends upon a rotating magnetic field to drag the shaft around, so the rotating magnetic field of the nascent sun continued to interact with the rings after they had been shed, tending to urge them ahead in their orbits as the contracting mass itself gathered speed. The force impelling the rings forward was accompanied by a braking force which gradually slowed the sun down, ultimately bringing it to its present one revolution in 25 days. In other words, the angular momentum gained by the rings came at the expense of the sun. These considerations have given new life to the nebular hypothesis.

Further evidence for planets comes from the observation of rotating stars. Since the time of Galileo we have known that our sun turns. Within a few days of building his first telescope, Galileo found that the sun had spots on it; the next day he found that the spots had moved and on the following day that they had moved again. From his original sketches of the sunspots, which we still have, he was able to work out the length of time it takes the sun to rotate on its axis and also to find out which way the pole of the sun's rotation was tilted. With stars which appear so small compared to the sun, observation is more difficult, but this turning has still been noticed.

According to a discovery of Otto Struve stars fall into two basic, sharply divided categories: those that are turning slowly (like our sun) and those turning rapidly. We now presume that the rapidly turning stars failed to be braked by outlying rings of matter which could be the precursors of planets and therefore probably do not have planets, whereas the slowly turning stars very likely owe their lack of angular momentum to the fact that they possess planetary systems. Of the stars observed, most of them, in fact about 90 percent, are turning slowly. The changing fortunes of the theories of planetary origin have thus produced the optimistic expectation that *planets are a commonplace by-product of star formation.*

Peter van de Kamp of Swarthmore College is diligently constructing a circumstantial case for the existence of planets. For several decades he has been observing Barnard's star, a close neighbor 6 light-years away. He finds that it wobbles across the sky and interprets this motion as indicating the presence of two invisible companions.

Perhaps future observation will completely surprise us and fail to reveal that many other stars have planets, but what follows is based on the general view of astronomers today that planets *are* a common by-product of star formation. We now focus on the conditions of the planets themselves.

Habitable Zones

Around each star is what is called a *habitable zone* (see *Fig.* 2) based on the idea that being too close to a star is too hot for life and too far away is too cold. Although it does not matter precisely what temperature limits are set, we consider that the habitable zone embraces the range of temperatures where water is liquid, being neither in the form of sterilizing steam nor permanently frozen solid. By the time conditions reach the temperature of steam, life doesn't seem too plausible. Some people assert that some forms of life might be able to exist in boiling steam. If there were such a thing, it would be a bonus over and above what is more likely to be the case. At the other end of the temperature scale where water is permanently frozen, again, life may be able to exist somewhere. But chemical reactions are slowed down so

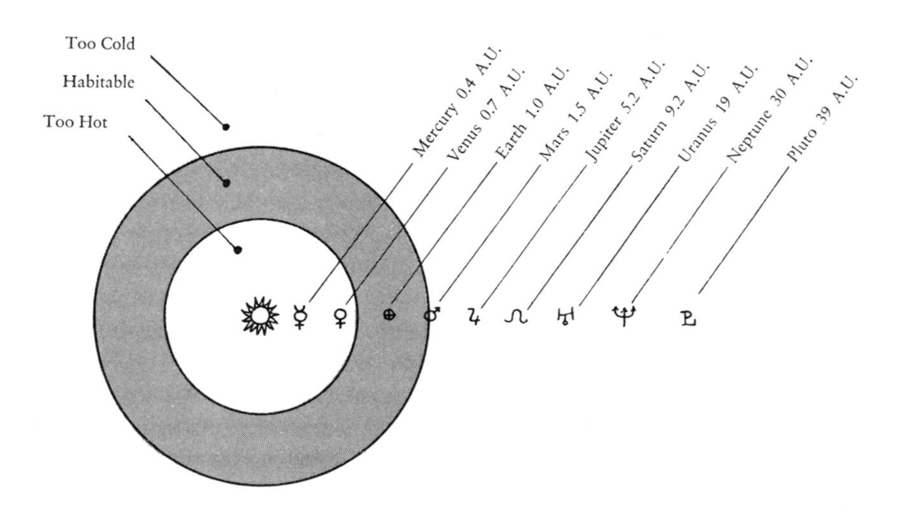

Fig 2. The earth lies in the habitable zone of our sun. Venus and Mercury are on the hot side of the habitable zone; Mars is on the cold fringe; the other planets well to the cold side. Distances from the sun are given in astronomical units. (1 A.U. = 93 million miles.)

much under these conditions that the possibilities of life are greatly hindered. With this idea of a habitable zone we can see that the cooler the star, the closer and narrower its habitable zone will be, and the less the probability that even one planet will be found to be in it.

We would also not expect to find life around hot stars because their life expectancy is short. Thus, O-type* stars, which are very massive, burn their nuclear fuel fast, radiate their energy away rapidly, and live for only about 10 million years as stable power sources. Since the geological ages on earth lasted for hundreds of millions of years, 10 million is probably not long enough for advanced forms of life to evolve. Perhaps exceptions exist, but the hot stars are not where we expect to find the most abundant life. Su-Shu Huang, who first assembled this line of discussion, concluded that life is to be sought around the slowly turning F-, G-, and K-type stars. Our sun is a G-type.

Not all slowly rotating stars classified in the vicinity of G-type are likely to be seats of life, however, for a variety of reasons. First, when examined closely, a good fraction of all stars prove to be double. Either two distinct stars can be seen in the telescope or, if a pair is too closely spaced to be distinguished separately, its true nature may be revealed by periodic eclipses, as one star gets in front of the other. A spectroscope, which splits starlight into colors (as Newton first did with sunlight), may reveal the to-and-fro motions as the components circle each other. A planetary orbit around a double star need not be an ellipse and is often irregular. Therefore a planet of a double star may not experience the strict constancy of solar heat that has characterized the development of life on earth.

We know that the sun's heat has been steady between very narrow limits for at least 400 million years, since there are corals of that age with living representatives today that can endure only very few degrees change in temperature without being killed. Thus if the earth had pursued a non-circular orbit that changed its average distance from the sun even very slowly, the chances of the evolution of the primitive early organisms would have been seriously impaired. While nature has shown amazing ingenuity in adapting to adverse circumstances and may have succeeded in establishing life in the environs of some double stars, in general they would not be the place to look, even if both component stars were of G-type.

A second handicap to the appearance of life would arise from the *size*

* Ranked in order from hot to cold the principal star types are O, B, A, F, G, K, and M. Since the days when stars were first classified and assigned letters of the alphabet there has been considerable revision. Additional types R, N, and S do not fit well into the sequence. Astronomers recall the order by reciting, "Oh, be a fine girl, kiss me," to which some add, "Right now, smack!"

of the planet that happened to be in the habitable zone. If it was too small, its gravity would not be sufficient to retain an atmosphere. The atmosphere would evaporate, just as hydrogen and helium have evaporated from the earth's atmosphere. In the case of our moon, all gas leaking out of the rocks evaporates. Without the benefit of the blanketing layer of air that the earth possesses, such a planet would be exposed to the lethal ultraviolet radiation and X rays that our ozonosphere and ionosphere absorb at heights of 30 kilometers and more. Conversely, if the planet was too massive (e.g., Jupiter), the sheer quantity of atmosphere would slow down the rate of evolution. It would take longer for the greater quantity of primordial methane and ammonia to be replaced by an atmosphere rich in oxygen, a commodity that most terrestrial life depends on for the generation of energy. The present healthy mixture of oxygen in the earth's atmosphere did not originally exist but was a byproduct of biological action.

Only about one in ten of the slowly rotating stars would be likely to support life after we reject double stars and planets whose atmospheres are too light or too dense. Remaining would be a total of about 10 billion suitable stars in our own galaxy. Of these, many probably have a planet of the right size in the right place. Although we do not know what this fraction might be, a conservative estimate of the total number of such planets in our galaxy would be around *one billion*. Whether life would originate, given the existence of a suitable planet, is not known. Nevertheless, given that one billion planets in our own galaxy may exist that are similar in habitability to our own earth, let us now consider the current state of knowledge about the origin of life.

The Origin of Life

Wonderful myths from all over the world attest to the universal preoccupation of man with the question of the origin of life. The key scientific explanation prevailing today dates from the time of Darwin in the 19th century. The notion of organic evolution introduced by Darwin and Wallace, if traced backward in time, brings us ultimately to the original common ancestor and suggests that the continuous development of each species from its immediate precursors may also apply to the development of the living from the nonliving. This idea was slow to take root, probably because the gap between the nonliving and the living is so great as perceived in everyday life, but it was taken up around 1930 by J.B.S. Haldane and A.I. Oparin. Both hit on the notion, soon to be supported by the discovery of methane and ammonia in Jupiter and Saturn, that the atmosphere of the earth was not always oxygenated as it is now. Oparin discussed the formation of organic molecules (such as the oils and waxes constituting petroleum deposits) in a hydrogen-rich atmosphere, while Haldane pointed out how the formation of organic molecules needed for life could more easily occur in the absence of atmospheric oxygen. During the course of time oxygen (liberated as a byproduct of life processes) altered the atmosphere, resulting in widespread changes. Iron, an abundant element that was previously rust-free, was oxidized, coloring the red rocks and sands seen all over the world. The conditions under which the inanimate spawned the animate do not exist here now and never will again. But the starting conditions are now perceived more clearly as the result of the work of Haldane and Oparin, and the great advances in molecular biology made since the 1930s have clarified chemical aspects of life—narrowing the gap between the nonliving and the living.

We are now looking for sequences of events on a molecular scale that lead to agglomerations that can reproduce themselves. A clue to the mode of reproduction has been furnished by the discovery of the structure of deoxyribonucleic acid (DNA), the genetic material that carries the code for reproduction of life. Throughout history it had been a total mystery how a single egg cell, seed, or spore could contain all the information for specifying everything about the new individual organism that would develop from it—all the instructions on how to assemble the parts and the means for procuring the necessary raw materials from the environment. James Watson and Francis Crick discovered that the DNA molecule is a double helix of two long, intimately meshed complementary templates. This double helix structure permits reproduction by peeling apart into halves and reforming on the template principle. (When one half peels itself off, a replica assembles itself from the surrounding nutrient medium with the aid of enzymes, until there are no openings left on the template. The openings on the template are tailored well enough to guarantee that the new assemblage is an exact replica of the previous one, even though it is very long and embodies much intricate detail.) This clever trick enables us to see how in principle the requisite amount of information can be packed into a cell, and many of the details are beginning to be discerned by molecular biologists. The template principle is surely the clue to guide the search for the origin of life.

Since it can be argued that the simplest molecule that can reproduce itself in a naturally occurring medium has life, a great search is now on. No doubt some sections of the life sequence will reveal themselves; one of these has to do with chemical evolution.

Chemical Evolution

If we begin by asking what the conditions on earth were before life appeared, we find a very different picture from what we see today. About 4 billion years ago (which is about a billion years after the formation of the earth) the earth's atmosphere did *not* consist mainly of nitrogen as it does today. Instead it contained primeval gases composing the nebula out of which the sun and planets were formed or gases exhaled from the body of the earth. In either case, the principal atoms of these gases were hydrogen, carbon, oxygen, and nitrogen—appearing in chemical combination as methane, ammonia, carbon dioxide, and water, or as related simple compounds. Under the action of cosmic rays from space, ultraviolet radiation, X rays from the sun, and terrestrial lightning discharges, the primordial composition was modified—and in ways that can be studied in the laboratory.

Since 1953, many scientists have tested various mixtures of the likely simple primeval gases, subjecting them to a striking assortment of agents: various electromagnetic radiations, high energy particles, electrical discharges. Always the results are the same: organic compounds are formed, many of which are immediately recognizable as key components of biological organisms. Sugars, fats, and amino acids (such as glycine and alanine) are among the products, as are long polypeptide molecules and ATP (adenosine triphosphate)—the substance used for energy storage by muscle tissue. Even wave action in water has been found to lead to the formation of organic molecules. Some of the more complicated molecules synthesized in the laboratory were made by energizing mixtures of the end products of earlier experiments together with naturally occurring compounds such as phosphoric acid. This is a reasonable stratagem for compressing millions of years of natural evolution into a few hours. The common molecular units of life have been produced in these tests. When we look back on the history of organic chemistry we recall that as late as 1828 it was a surprise when Wöhler, the founder of organic chemistry, announced that he had synthesized urea (the first organic molecule to be made in the laboratory) "without the help of a man, a dog, or a kidney." We have now found that the inanimate world can synthesize organic substances *unaided.*

The Primordial Soup

These experiments usher in that aspect of chemical evolution in which we attempt to determine which substances were produced before life appeared, in what quantities, and what further reactions would take place, for example in the sea. From the starting point of just over 4 billion years ago a progressive accumulation of organic matter must have

taken place in the seas. In shallow seas or in lakes subject to strong evaporation the concentration would be higher, thus bringing together the right molecules at the right time and allowing the formation of a more elaborate compound that might be an essential ingredient in the next step. Some of these next steps would be similar to chain reactions causing rapid transformation of the contents of the lake or mudpool.

Picture a small lake, just over 4 billion years ago, filled by rains that bring down atmospheric constituents, such as ammonia, that make the lake water rather different from what we have today. The solar ultraviolet rays beating down upon it form the sugars, amino acids, and other compounds found in the laboratory simulations. Elements leached from the rocks, such as calcium, potassium, and phosphorus, add themselves to the brew producing further, more complex reactions. Season upon season in lake after lake, the mixture failed to reach the ability to reproduce or even conserve itself before it either evaporated or was washed away. With the passage of time, an increasing abundance of such substances accumulated, until at some time and place the threshold was passed—and reproduction began, probably in a small lake.

A lake that evaporates to a small volume may experience rapid production of end products that would appear only in traces in the ocean, both because its contents become concentrated, permitting fast reaction rates, and because of chain reactions. A chain reaction facilitates its own continuance by the formation of intermediate products that in turn react with the original ingredients. (A well-known example is the fission of uranium.) A sufficient initial concentration is necessary in order to initiate a chain reaction, whose outcome is to transform the whole supply of an original substance into another. A small body of water undergoing periodic evaporation offers such a possibility.

With regard to the atmosphere, the outcome of chemical evolution was to extract some of the carbon from the carbon dioxide and incorporate it into living organisms or into their remains (coal and limestone). Oxygen was liberated from water or minerals; the hydrogen in the methane and ammonia found its way into the form of water or was lost by evaporation from the top of the atmosphere; the residual nitrogen and oxygen are what remain today.

Time upon time particular lakes must have manufactured batches of diverse and interesting chemicals in quantity without hitting on the sequence of steps needed for self-reproduction. We hope that continued progress in the understanding of the evolution of our atmosphere, aided by knowledge of atmospheres of other planets revealed by astronomical observations and the space program, will combine with advances in molecular biology to reveal *what* these steps might have been.

Early Life

More than a billion years elapsed as the earth cooled and the precursors of life assembled themselves in the primordial soup before unicellular organisms as known today began to appear in the fossil record. Such organisms may, of course, have developed earlier but without leaving any trace—or their traces may still be awaiting discovery. The earliest now known are single-celled blue-green algae which occur as fossils in ancient rocks and are approximately 3.5 billion years old. They continue to flourish today in ponds. Although the blue-green algae are primitive, they possess a cell wall, nuclear material, and provision for photosynthesis. They represent a considerable evolutionary step beyond the bacteria and viruses which seem likely to have preceded them in the sequence of development. The algae do not feed on byproducts of other life but live directly on minerals leached by rain from the rocks (namely metal phosphates and nitrates) and depend on sunlight captured with the aid of chlorophyll for their energy. As these simple requirements are available today, the blue-green algae are still with us.

A key question remains unanswered: Was the evolution of primitive life a more or less inevitable consequence of the conditions it arose from; or, was there an episode whose chances of occurrence were so low that it is unlikely that, if the clock were turned back, life would again appear? One might be inclined toward the latter thought on the grounds that a billion or more years of chemical evolution is a long time to wait for something to occur that is inevitable. Yet, since we do not know all the steps, but only that they were probably very numerous, the time scale might be perfectly appropriate for their occurrence one by one.

Accident or Destiny?

The *rare episode* theory would have to be explained in far more scientific detail before I could be persuaded of its reality. This episode would have to be so crucial that its failure to happen would constitute *a total blockage to all possible paths leading to life.* Nature excels in her impressive ability to adapt, by random and repeated trial, to the circumstances she is confronted with. Consequently, as a basis for discussion, I assume that self-propagating life is a thing to be expected, given time and suitable circumstances. But I would not necessarily expect the blue-green algae.

Undoubtedly the availability of water is one of the suitable circumstances facilitating the propagation of life as it permits rapid chemical reaction, thus speeding up the time scale of evolutionary chemical events. The availability of carbon, permitting the rich chemistry of mo-

"Ammonia! Ammonia!"

lecules containing carbon (unequaled by any other element), is another definite plus. However, one would have to be prepared for the possibility that somewhere radically different organisms might exist, not based on organic chemistry (the chemistry of the carbon atom). The element silicon is often mentioned as a conceivable basis for forms of life that would be profoundly different from ours, because it combines with other chemical elements in much the same way that carbon does. But it is not easy to envisage silicon-based life. Hydrocarbons such as methane, when oxidized, have end products of water and carbon dioxide, the latter exhaled by earthly animals. However, silicane oxidizes to water and silicon dioxide, which is quartz, or in granulated form sand, and would be hard to breathe out. I am not saying it is impossible that there are creatures somewhere that breathe abrasive grit out through stone nostrils. While not expecting them to be a substantial fraction of all life forms, we may nevertheless accept them as a bonus.

Whether it would be conceivable, in an initial environment such as the earth's, to have evolution of life *not* depending on DNA is an intriguing question. Is this basic plan such a superior and inevitable property for life that a competitor is unthinkable? If life is found in space will it utilize the same plan for genetic coding as all life on earth does? If, sometime in the past, living organisms on earth existed tracing back to other origins, they were consumed.

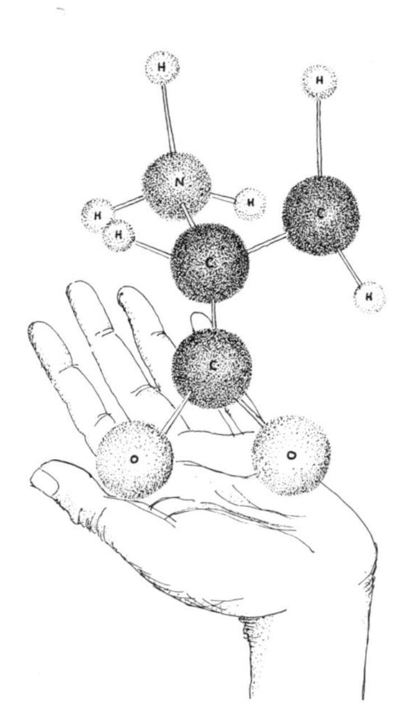

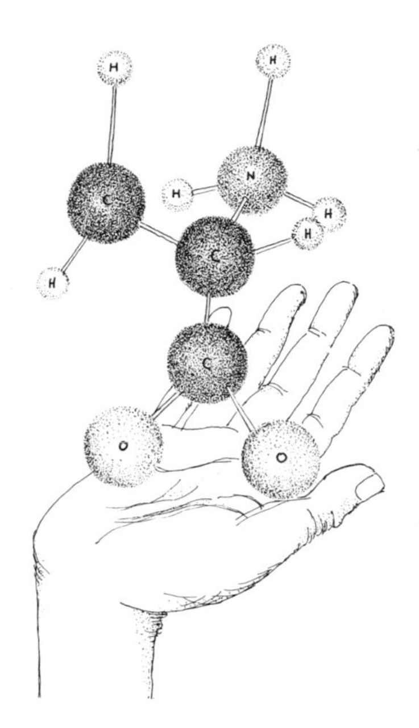

Complicated molecules may exist in opposite-handed forms, which have the same relationship to each other as the gloves of a pair, one being the mirror image of the other. But an amino acid occurring in the protein of a living organism tends to be of the one form, not of the other. For example, in natural protein we find only *L*-alanine, the form on the left-hand side of *Fig. 3*. Perhaps this is because the original jump from the nonliving set the pattern which has been followed ever since. Or, it is possible that the environment possessed a built-in bias and, over a long period of natural selection, favored one variety of molecule at the expense of the other. It is, however, conceivable that mirror-image life flourished at one time. If life is found in space, will the molecular byproducts of life be the same as or mirror images of those on earth? Assuming that living matter can one day be synthesized in the test tube, will both forms of molecule prove to be viable? What would happen if mirror-image molecular life was released into our environment? This is of fundamental importance. When the surface of Mars comes to be sampled by the Viking lander in 1976, the molecules will be tested for chirality (handedness) as a sign of biological activity.

Fascinating questions are raised by the possibility that we are not alone in the universe, many of them to be taken up in later chapters. But let us be clear: *To date there is no evidence commanding wide assent that there is life in the universe elsewhere than in the earth-moon system.* For all we know, earthly life may be the only life there is. I personally judge it to be more likely than not that there is intelligent life elsewhere in our universe, outside our solar system. I do not exclude the possibility that we are the only intelligent beings; however, if I adopt that assumption there is really nothing more to say, whereas the opposite view leads to interesting trains of thought.

CHAPTER TWO

VELIKOVSKIAN VERMIN

SIMPLICIO *(quoting a classical author): "If the earth were to stop by the will of God, would other things continue to rotate or not? . . . it would be very remarkable if the seagull could not hover over the small fish, the skylark over her nest, or the crow over a snail or a crag, though wishing to do so.*

SALVIATI: *For my part I should give a general answer: That if by God's will the earth should stop its diurnal whirling, the birds would do whatever that same will of God desired."*

Galileo, *Dialogue Concerning the Two Chief World Systems*

IN HIS WELL-KNOWN BOOK *Worlds in Collision* (1950), Dr. Immanuel Velikovsky discusses the possibility of low forms of life inhabiting Jupiter and Venus. Instead of cautiously using a neutral term such as "living organism" he uses the word "vermin," which has had a provocative effect on some of his scientifically inclined readers, calling forth some scorn and derision. He also entertains the notion that this life was transported to earth by a comet. Velikovsky writes:

> The old question, whether there is life on other planets, has been debated time and again without much progress. Atmospheric and thermal conditions are so different on other planets that it seems incredible that the same forms of life exist there as on the earth; on the other hand, it is wrong to conclude that there is no life on them at all.
>
> Modern biologists toy with the idea that microorganisms arrive on the earth from interstellar spaces, carried by the pressure of light. Hence, the idea of the arrival of living organisms from interplanetary spaces is not new. Whether there is truth in this supposition of larval contamination of the earth is anyone's guess. The ability of many small insects and their larvae to endure great cold and heat and to live in an atmosphere devoid of oxygen renders not entirely improbable the hypothesis that Venus (and also Jupiter, from which Venus sprang) may be populated by vermin. (Dell, p. 194.)

Conjectures about special forms of life, different from those on earth, are supported by some well-known astronomers, among them Dr. Carl Sagan, the Cornell astrophysicist. Sagan has discussed some of the satellites of the solar system as possible life-supporting habitats. In 1966 in *Intelligent Life in the Universe* he wrote:

> It has been customary to dismiss instantly the possibility of life on Jupiter, with a reference to poisonous gases and freezing temperatures. But the gases of the Jovian atmosphere, let us recall, are far from unambiguously poisonous; indeed they are just the components of the primitive atmosphere in which life arose on Earth. (Holden-Day, p. 328.)

Although Jupiter's cloud tops, which are what we see through a telescope, are very cold, the temperature is warmer inside the clouds. There may even be a thick cloud of liquid water droplets. Sagan continues:

> It therefore seems inescapable that large quantities of organic molecules are being produced abiologically in the atmosphere of Jupiter today, and that such conditions have been maintained for the past 4.5×10^9 years. Jupiter is in fact an immense planetary laboratory in prebiological organic synthesis.
>
> It is much more difficult to say anything about the possibility of the origin and present existence of life on Jupiter. For example, we can imagine organisms in the form of ballasted gas bags, floating from level to level in the Jovian atmosphere, and incorporating pre-formed organic matter, much like plankton-eating whales of the terrestrial oceans. (p. 329.)

According to Velikovsky, the route by which life on Jupiter might have been transported to earth is related to the way in which he believes the solar system developed, a mode of development conflicting with the nebular hypothesis. If correct, Velikovsky's theory would significantly alter our understanding of the appearance of life both on earth and on other planetary systems in space. What follows is a synopsis of events that Velikovsky deduces from various sorts of historical evidence.

Venus Sprang from Jupiter?

Velikovsky concludes from studies of ancient literature that Jupiter expelled a comet which touched the earth's atmosphere around 1500 B.C. causing worldwide catastrophes, brushed the atmosphere again 50 years later, and after orbital changes settled into a circular orbit around the sun, where we now recognize it as the planet Venus. The first encounter between this comet and the earth supposedly took place during the departure of the Children of Israel from Egypt, causing irruptions of the sea that destroyed the Pharaoh and his men and brought an end to the Middle Kingdom. Two months later, when Moses received his revelations on Mount Sinai, earthquakes and volcanic activity caused by the world-shattering impact were still in progress. The Israelites that escaped wandered in the desert for 40 years under a deep gloom of smoke and dust that blotted out the sun, surviving with the aid of manna, a "sugar" that fell in grains from heaven. During this time a series of plagues afflicted the land. Creatures such as flies, frogs, caterpillars, locusts, lice, and serpents are mentioned in Exodus, where they are referred to as vermin. According to Velikovsky, these happenings were caused by the comet that was soon to become Venus. The return of the comet about 50 years later coincided, he says, with the earth tremors and hail of meteorites that occurred at the time of Joshua's battle—when the sun stood still. Worldwide devastation was caused by the surge of waters when, according to Velikovsky, the earth's rotation was interrupted—changing the latitudes of places on earth. This facet of Velikovsky's rendition, set forth here in only the briefest abstract, as well as others (e.g., a collision between Mars and earth), have proven too much for many people to swallow, particularly experts in astronomy, geology, Biblical studies, assorted ancient literatures, and archeology. Attacks on his work have generally been scientific criticism of particular claims or a political discrediting of his whole work.

A Scientific View: The Orbit of Venus

In Velikovsky's scenario, one of the things that immediately strikes an astronomer is how difficult it would be for an interplanetary body, such as a comet, to pass from one orbit to another. This is especially difficult if, as in the case of Venus, the orbit is highly circular. There is a great difference in the energy associated with different orbits and, in order for the excess energy to be eliminated, something drastic would have to occur. The cataclysmic events proposed by Velikovsky are a collision of the comet with both the earth and Mars, as well as an earth-Mars encounter. Yet, instead of all three bodies lodging in the highly circular orbits they occupy today, they would more logically be moving in intersecting elliptical orbits. Velikovsky does not supply an explanation for Venus having a circular orbit and does not refer at all to the extraordinary degree of circularity of the orbit (the major diameter is 216,400,000 kilometers and the minor diameter is a mere 5,000 kilometers less). Instead, he contents himself with pointing out that the circular orbits of the planets and their satellites are in any event a mystery. This may have been the case when he wrote in 1950.

Until the 1950s, magnetic forces were not invoked in attempts to understand the solar system. (Velikovsky himself pointed out that magnetic forces should be included in celestial mechanics.) It is now universally agreed that magnetic forces influence the behavior of the sun and formerly influenced the nebula from which the sun and planets condensed. Magnetic forces must have been at work if Venus was once a ring of coalescing matter that underwent slow orbital evolution. Magnetic and frictional forces offer hope of explaining a rounding off to today's orbit. But no mechanism has been proposed that could convert a *comet* to a *planet in circular orbit* in a few thousand years! That orbits evolve has become apparent from the satellites of Jupiter and Saturn, whose orbital periods have progressively changed only to become locked at distinctive values. One cause of orbital change is internal friction; another is the friction due to a planet plowing through the remnants of interplanetary matter left from the nebula from which the planets condensed. Although over 1,000 tons of meteoritic matter from interplanetary space still falls on the earth each day, primarily as dust, the net force applied today is small. But accumulated interaction with the interplanetary medium has shaped the planetary orbits and in the case of Venus has synchronized the Venus day with the earth year.

In time, the history of the planetary system will be pieced together, but I will be surprised if the record reveals that Mars, Venus, and the earth were rattling around like dice in a box in 1500 B.C.

Changing Latitudes: Continental Drift

Another of Velikovsky's claims is that places on the earth have changed latitude. To support this, Velikovsky refers to geological evidence such as the discovery of fossils of tropical plants in arctic regions. To show that drastic changes occurred in historical times, he quotes extensively from literary sources which he interprets as indicating that the rotation axis of the earth shifted at the time "the sun stood still."

Subsequent to the time of Velikovsky's writing (1950) an explanation for these geological findings has become generally accepted—the understanding of continental drift. As the earth formed by a coalescence of interplanetary matter, it was not at first perfectly round (nor is it yet). Several hundred million years ago it was slightly pear-shaped. Although more than half was covered by ocean, as it still is, the top portion of this pear was covered by a single landmass. As gravity works to round the shape of the earth, the next step in the approach to roundness was a splitting of the landmass into continents which drifted apart. Continental drift explains how fossils were carried from the tropics to colder latitudes, thus eliminating the geological support for Velikovsky's conclusion—that Venus jolted the earth, carrying life from Jupiter, and caused a change in the earth's rotational axis.

Was Venus a Comet?

I have not yet dealt with the evidence that Velikovsky adduces for thinking that Venus was once a comet. A collection of historical references purports to show that Venus once had a tail and was much brighter than it is now. These references range from pre-Columbian Mexico, Babylon, and the Ganges valley, to Egypt and China, and include Hebrew writings. Although a substantial fraction of the quotations presented could apply to a bright comet, they are also applicable to other striking celestial appearances, such as Venus as it appears today, or to a conjunction of Venus with some other phenomenon such as a bright meteor or an atmospheric halo. Venus itself is a splendid object, quite capable of casting a shadow at night, and of being seen in broad daylight with the naked eye. Few city dwellers of today are aware of the brilliance of Venus from personal observation.

A Cometary Encounter?

Perhaps the foregoing discussion will help the reader to form an opinion about the likelihood that Venus was expelled in the form of a comet from Jupiter, and, after colliding with the earth and Mars, came to rest in its present orbit—slowly changing into a planet. A word of caution may be in order. Merely because Velikovsky's grand synthesis is unconvincing at some points, it does not follow that everything he says is wrong. One of the charms of *Worlds in Collision* is the rich abundance of assorted information that the author offers. It is indeed possible that the events around 1500 B.C. were produced by the earth encountering a comet.

No clear observational experience exists for judging what might happen if the earth passed through a comet. However, we have experience with meteors, and the connection between meteors and comets is illustrated by the story of Biela's Comet, which used to return every 6.6 years. In 1845 two comets appeared rather than one, and in 1852 both reappeared—but not so close together in the sky. Evidently the nucleus of the comet had split, and the parts were separating. Neither part was ever seen again. But subsequently, meteors have been seen when the earth crosses the former comet's orbit.

If the earth passed through a comet many meteors would be seen, and the larger ones would be followed by meteorites striking the ground. A famous meteorite shower occurred at Laigle, France, on August 26, 1803, when about 3,000 fragments fell in an elliptical area 6 miles long and half as wide. On January 30, 1868, an estimated 100,000 fragments fell on an even smaller area at Pultusk, Poland. In a comet's tail, we would expect to find objects large enough to form craters (similar to lunar craters, which are essentially explosion pits).

The Arizona meteor crater was formed by a large mass of iron, estimated to be 20 meters in diameter and to weigh 50,000 tons, that blew a hole 1,240 meters across and 190 meters deep. At the time a tremendous blast wave must have devastated the area and a great plume of dirt must have rocketed into the air producing a fallout of dust that continued for some time—no doubt darkening the surrounding sky. There are no eyewitness accounts of this encounter, but a Siberian meteoric explosion of June 30, 1908, was seen and heard hundreds of kilometers away. A jet of flame shot miles into the air, the surrounding forest was *flattened* out to a 30-kilometer radius, and seismographs recorded tremors all around the world. Rivers flooded, houses collapsed, 200 kilometers away horses were blown over, and 700 kilometers away horses could not stand up. The night sky was brightened all over Europe by noctilucent clouds, possibly by coincidence, and the intensity of sunlight was diminished for several days in faraway California. In 1946 a meteoric explosion in Kenya knocked down huts for a distance of 50 kilometers, a village was burned and its cattle destroyed by a hail of meteorites.

Compared to the effects of the Arizona meteorite, some 20 meters in diameter, the solid nucleus of a comet (about a kilometer in diameter) would produce a world-shattering impact on a scale well beyond the largest H-bomb explosions now available to the hand of man.

The Vermin Multiply

It does not seem implausible to me that a cometary encounter could have taken place in historical times, and that earth-wrenching tidal

forces could have triggered seismic and volcanic activity as well as great tidal waves and floods. Forty years of darkened skies due to continued volcanic eruption and direct meteoritic deposition seems possible. Dust belched out by just one eruption of Indonesia's Mt. Tambora in 1815 so obscured sunlight as to bring record cold as far away as Europe, where the English summer a year later was the coldest recorded in 250 years. If protracted obscuration occurred, vegetation would be seriously affected and the food chain for both vegetarian and carnivorous creatures would be disrupted. One can imagine the wolves devouring the last of the rabbits as they emerged in search of the last withering herbage. After that, as carcasses provided food for scavengers, they would multiply, especially the multitudinous flies, beetles, and other invertebrates whose function in the scheme of things is to clean up—in other words, Velikovsky's "vermin." Nocturnal creatures such as cockroaches would appear in the day and, on the whole, the picture conjured up by Velikovsky from Exodus and other sources is a conceivable natural consequence of obscured skies that never seemed to clear. But there must have been brighter interludes sufficient to keep life going.

No evidence has been presented of the disappearance of trees and plants known before 1500 B.C., but a report of 40 gloomy years could have resulted from repeated outbreaks of volcanoes, with more or less clear intervals between. Independent evidence for vulcanism at that time includes the violent eruption and virtual demolition of the Greek island of Thera (Santorini in Italian) which wiped out Minoan culture. The volcanic activity of the Mediterranean is connected with movements of the earth's crust that might well be triggered by a cometary encounter.

Literary investigations along Velikovsky's lines, supplemented by geological field studies and radiocarbon dating, should establish what unusual global phenomenon, if there was one, occurred around 1500 B.C. Curiously, Velikovsky himself pushed hard for radiocarbon dates only to find inertia in museums possessing suitable material. According to correspondence reproduced in *Pensée* (vol. 4, no. 1, 1973), the British Museum appears to have discarded, or denied obtaining, results that were inconsistent with preconceived chronology of the Egyptian dynasties. Clearcut tests of Velikovsky's ideas seem to have been avoided on the grounds that the ancient materials used were contaminated. Yet, we can see how the early inconsistencies thrown up by the new radiocarbon dating technique could cast suspicion on the materials themselves. We now know that a radiocarbon date depends on the intensity of cosmic rays at the time and that cosmic rays have varied in intensity over historical times, distorting the time scale. Thus radiocarbon dates are subject to correction. This discovery came from age determinations on bristlecone pine trees, the world's oldest living things, made by counting annual growth rings in the wood and then comparing this with the age determined by radiocarbon. As cosmic-ray intensity has had a worldwide effect, the ancient trees of California have permitted significant revision of archeological chronology in Europe. This tool for obtaining comparative dates for geographically separated cultures will ultimately confirm or dispose of the surmised encounter of 1500 B.C. with its worldwide floods and disasters.

Life from Jupiter

As to the possible truth of Velikovsky's assertion that life from Venus or Jupiter may have been transported to the earth by the supposed comet, chances are that any outside life would have been consumed by existing terrestrial organisms. If the alien life had established a foothold, and was still with us, it would possess its own peculiarities. All living organisms share certain complicated protein molecules; even organisms as divergent as plants and animals share the same proteins in their most central process: reproduction. This is generally interpreted to mean that all terrestrial life has a common origin, animals being descended from single-celled animalcules that in many ways are not very different from plants.

Any alien creature not sharing the molecular history of those on earth would clearly announce its foreign origin. But no such creature has been found. Velikovsky's notion that terrestrial life may include organisms brought from Jupiter does not seem likely to influence biology. It would certainly electrify biologists if, by accident, evidence for Jovian organisms were discovered on earth—I, however, do not expect this to occur.

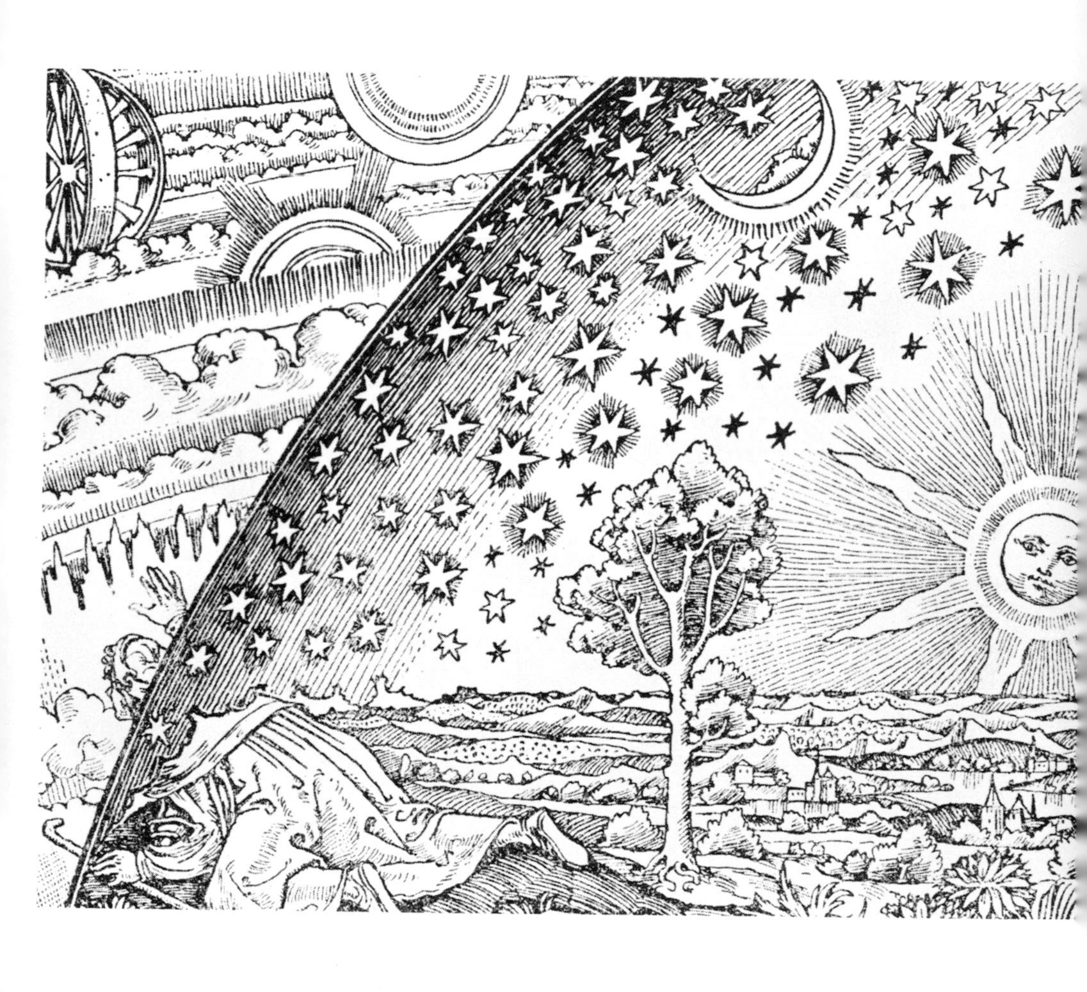

CHAPTER THREE

IS MAN UNIQUE?

Strange events permit themselves the luxury of occurring.

Charlie Chan

HISTORY TELLS THE STORY of human chauvinism. Throughout the ages when man has assigned himself preeminence, expanding knowledge has often proved his assumptions wrong. The Chinese civilization regarded itself as being in the center of the world, as we see from the word *China*:

中國

The sign on the right means *country* and the other signifies *middle*. We should find such chauvinism familiar, given the heritage of Western civilization; classical Rome and Greece felt the same way about themselves. The word *Mediterranean* means middle of the earth; the Greek word *Mesogeios* is the equivalent. As things turned out, the earth proved not to have four corners, denying any country claim to the center.

Another example of hubris was the Aristotelian belief that the earth was the center of the universe, a theory used by astronomers for centuries to explain the motion of the planets against the starry background, until displaced by Copernicus, who assigned the place of honor to the sun—*our sun*. Only later did we learn that the sun is but *one* of about 100 billion stars arranged in a single galaxy which reveals itself to our eyes as the Milky Way. The sun is neither in the center of our own galaxy nor remarkable in any way. (More recently, we have become aware of a universe of over a billion galaxies.)

If there are other intelligent communities in space, some will be more advanced than ours and some less. When we bear in mind the accelerating pace of technology and science, we see that those who are ahead of us will be very much ahead of us in their understanding of and control over their physical environment. Western civilization of 20 years ago was impressive compared with that of primitive areas, but perhaps not much more so than Roman civilization had been. By comparison, enormous strides have taken place in recent centuries and even greater ones in recent decades. (I am referring to man's understanding and control of the physical world and not to matters of morals and politics.) No one can foresee what will be achieved a century hence. And yet there may be, somewhere in the universe, communities that passed through our level of technology—with chemical fertilizers, electric power, radio, bombs, satellites, computers, molecular biology—a million years ago. Try to picture them today.

Could the Impossible Happen?

To suggest that we are the only intelligent beings in the galaxy would be, judging from most of the writings on extraterrestrial life, to suggest the *impossible*. However, some sobering thoughts have been advanced by Professor W.H. McCrea of Royal Holloway College, London. He points out that our nearest neighbors, Venus and Mars, despite their similarity to the earth in size and mass, have atmospheres which are radically unlike the one that has favored earthly life. The gross differences that exist today may have resulted from chemical compositions that were originally much the same but which were influenced by the unequal periods of rotation. A stroke of luck may have been responsible for our clement conditions. Furthermore, he notes that the western margins of North and South America have undergone great and rapid upheavals in recent geological ages and that if such instability were more widespread it would be very detrimental to advanced life. Surmising that the mass of the earth is just right, he concludes that the earth benefited from another stroke of luck.

Listing several such coincidences, he mentions that it might be necessary to have a tide-producing moon* in order to create the right condi-

* If there were no moon, tides would still be raised by the sun, but their range would be reduced to 1/3.

tions for getting life out of the seas and onto dry land. (The underlying premise is that the rich life forms of the intertidal zone have the ability to withstand drying out twice a day when the tide is out and are therefore in a good position to develop organs permitting them to remain out of water even longer.) He concludes that even with 1 billion planetary systems in the galaxy, and even with most of them producing life of some kind, an extraordinary conjunction of favorable omens would still be required to lead to *intelligent* life.

The orthodox(!) response to such gloomy comment is to insist that somehow, somewhere, that which has survival value *will* evolve, and that intelligence *does* have survival value. Still, without estimates of the rate of evolution to show that the supposed evolution can occur in the available time, one cannot be completely convinced. The case of the humble but venerable turtle, who traces back some 200 million years into the Mesozoic era, reminds us that intelligence is not the only factor contributing to survival; many creatures do quite well without it. Until a superior community is actually discovered, it is necessary to bear in mind that *the impossible may have happened* and that we may indeed be unique in the galaxy. After all, we are unique on earth.

Preempting Natural Evolution

It is clear that, at present, no species is emerging as an intelligent rival to man in any part of the earth—nor can one in the future, as long as man predominates. Man has now covered the planet, virtually precluding the possibility of independent emergence.

Likewise, even if given another billion years, life could evolve on Mars, it would probably not get the chance. Long before then Mars will be exploited by man (if we survive), blocking natural evolution. Of course, in one sense, the arrival of human life on Mars would be a consequence of natural evolution because, as man is a manifestation of nature, the migration of man to Mars would be merely a step beyond his expansion over the continents.

The point is that man has been able to diffuse from his point of origin to other continents, and soon will reach Mars, in a much shorter time than it would take for the whole chancy process of origin to repeat itself. This analogy applied to the galaxy as a whole is clear: *Will the time required for terrestrial life to diffuse through the galaxy prove to be shorter than the time necessary for new intelligent life to originate independently?*

The answer may be yes, in which case man would be a key feature in nature's plan to populate the galaxy. As yet, we cannot confirm or reject this possibility of human diffusion. As a representative speed for comfortable space travel we can use 1/10,000 of the speed of light, a conservative speed. Life could cross the 30,000 light-years to the center of our galaxy in 300 million years. Although this seems a long time, it has taken the earth 5 billion years to produce man. It is certainly not impossible that man's descendants (whatever form they may take and whatever speed of travel they may be able to achieve in the next 300 million years) will populate the galaxy and preempt the development of independent intelligent life which might otherwise evolve. This living galactic wave of our descendants advancing through space may encounter and engulf minor pockets of other life here and there. Conversely, perhaps the grandiose vision developed above will prove immodest, and the destiny awaiting us is merely to be engulfed by a greater civilization which is now expanding toward us!

In the absence of evidence to the contrary, we must consider the possibility that there may be intelligent societies elsewhere, some surpassing our level of development. We do not know in what direction or how far away our closest superior neighbor may be. How would we make contact with it? Will it contact us?

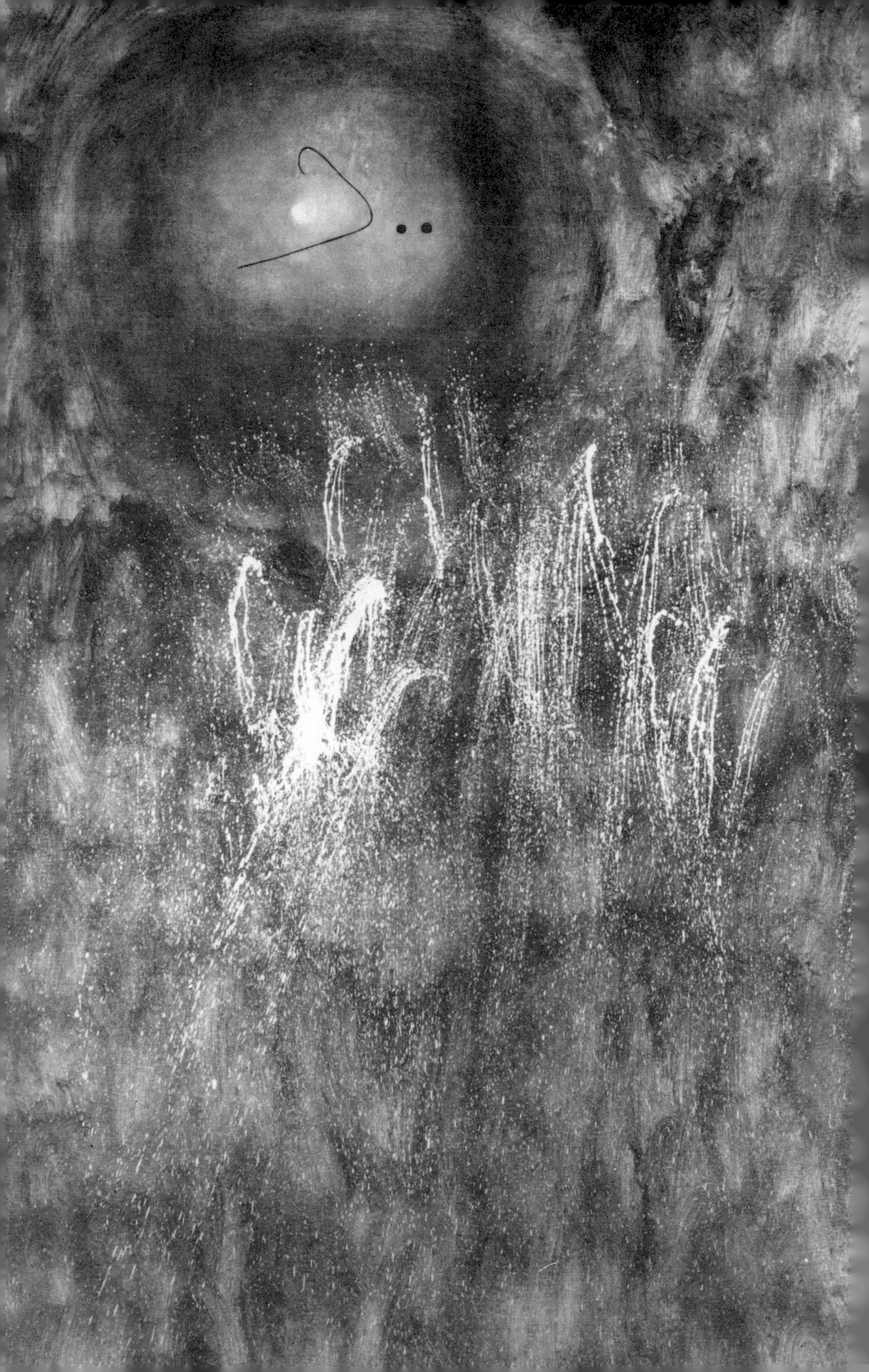

CHAPTER FOUR

SHOULD WE TALK TO THEM? AND HOW . . .?

For God's sake let us not answer.

Ždenek Kopal

SOME PEOPLE WORRY that it would be dangerous for us to encourage contact with intelligent extraterrestrials on the ground that the natives who ran out on the beach to greet the European seafarers often suffered. The visitors might carry us away and breed us for beef cattle, they say, or steal our mineral resources. But I think that it would be cheaper for them to synthesize steak chemically on their own planet and to replenish depleted minerals from uninhabited worlds closer to home. The cost and difficulty of interstellar voyages are such as to invalidate simple comparison with the contacts between terrestrial civilizations.

The chief commodity of interstellar trade will be information. It would be nice to have an artifact from another planet, a piece of art, or perhaps some seeds, but we will be able to get their poetry and music by radio and I suppose even sculpture could be reduced to a digital code that would permit reconstitution at the far end. I don't know whether color television will improve to the point where paintings can be transmitted faithfully, but perhaps the other party will not even be interested. They will be interested, however, in the hard knowledge that we have accumulated, even if they are far more advanced than we are. After all, we make expeditions to ancient sites of human habitation and painstakingly piece together the scanty evidence to form a picture of life at the time. No matter how primitive the inhabitants may have been, if we could revive one he could help us immeasurably. We have accumulated considerable information about our earth and solar system which would make contact with us very much more attractive than exploration of an uninhabited planet where such detailed data would be impossible to acquire.

Whose Initiative?

Assuming then that it is desirable to make contact with our more advanced neighbors, how should we go about it? *It seems certain that the party that takes the initiative in a project like this is the one with the superior technology. After all, Columbus discovered America; the American natives did not discover Europe.* Therefore, instead of sending signals we should be looking for the signals that may be beamed at us. A good deal of thought has been devoted to the method of signalling that might be adopted by a community desiring to attract our attention; powerful light beams from lasers have been suggested but because there are weighty arguments favoring microwave radio, it is worthwhile to discuss the radio telescope and terminology that will be used.

Fig. 4. This Stanford University radio telescope has a beamwidth of only 1 percent of the moon's diameter. Although it is designed for radio astronomy, it could be trained on distant stars to receive messages from immense distances.

The Radio Telescope

This picture (*Fig. 4*) shows a compound radio telescope at Stanford University comprising five interconnected paraboloidal reflectors, or dishes, as we call them for brevity. The paraboloidal shape, on intercepting rays from a distant source, redirects them all to a single point known as the focus, where an open-ended pipe collects the energy intercepted by the reflector and conducts it to a radio receiver, as shown in *Fig. 5*.

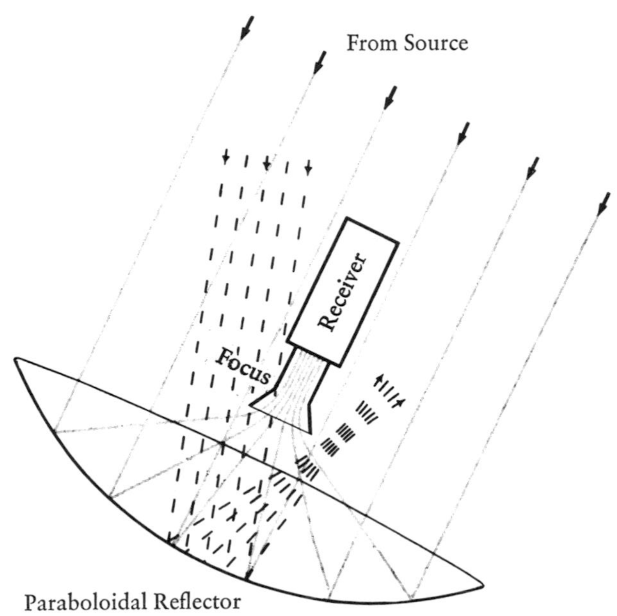

Fig. 5. Parallel rays from a distant source reflect from surface of paraboloid to come together at the focus where they are collected and passed to a radio receiver.

Directivity

A major purpose of using a paraboloid (or dish) is to enable us to receive very faint signals by gathering as much energy as possible and concentrating it. A second purpose concerns *directivity*, the ability to receive signals from one direction while rejecting signals from other directions. Thus if a second source of signals happens to be present, as indicated by the dashed lines in *Fig. 5*, these rays reflected from the dish will not be directed to the focus and consequently will not reach the receiver. Other examples of directive antennas are the elaborate tubular structures mounted on the roofs of some houses for their TV sets. These antennas are pointed toward the TV transmitters when they are installed, or may be provided with a rotator to change the direction in which the antenna is pointed. Most radio-telescope reflectors are mounted so that they may be pointed to any direction in the sky.

A different type of antenna, which is also germane, is encountered on the tall towers housing TV transmitters. The object is to radiate the power from the radio transmitter uniformly toward all points of the compass, that is, omnidirectionally in the horizontal plane. There is no need for power to go upward; ways have been found to confine the radio emissions toward the horizon, which is where the customers are (as seen from the transmitter tower). *Fig. 6* shows that some of the radio energy is not intercepted by listeners (the domestic antennas) and spills out into space. Also shown in the figure is the radio shadow or zone where radio transmission is obstructed, both beyond the horizon where the bulge of the earth intervenes, and in the lee of mountains, high hills, or buildings.

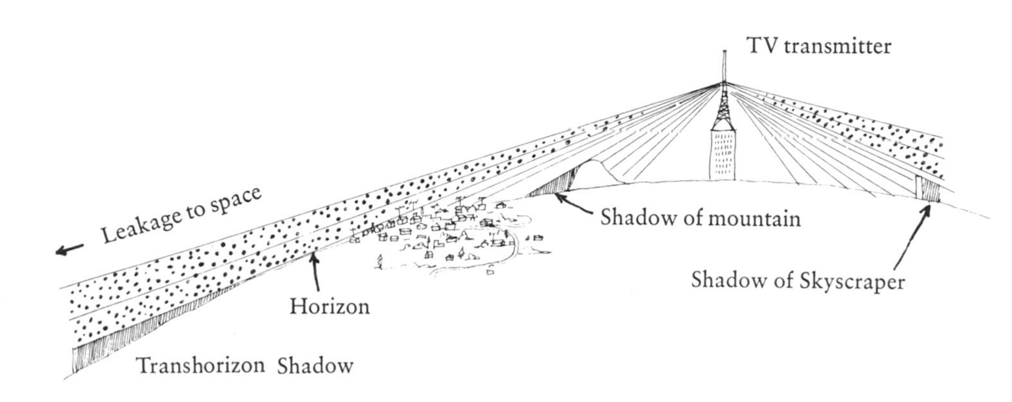

Fig. 6. Power radiated from a TV transmitter on a tall tower is deliberately concentrated toward the horizon in order to avoid wasteful radiation upward, but a thin sheet (dotted area) nevertheless escapes into space as leakage radiation.

Beamwidth

Any given antenna may be used either for transmitting or receiving and will have the same directivity whichever way it is used. A common way of expressing directivity is in terms of the *beamwidth*. Thus the beamwidth of one of the radio telescopes in *Fig. 4* is expressed as *7 minutes of arc*, meaning that transmitted power would be sent out mainly within a cone having a 7-minute angle. This is quite a small angle: 30 miles away the transmitted beam would cover an area only 100 yards across. Conversely, when the antenna is used to receive, it is capable of accepting a source 30 miles away while rejecting another only 100 yards to the side of it. Even greater directivities (or narrow beamwidths) are commonplace in radio astronomy: the *compound system* at Stanford University (*Fig. 4*) has a beamwidth of 18 seconds of arc, which is about 20 times narrower and would reject sources only 100 yards to the side at a distance of 600 miles. Turning now to the receiver, we may regard it as an amplifier characterized by three main attributes: 1) *frequency*, 2) *bandwidth*, and 3) *sensitivity*.

Frequency

The frequency, which specifies the point in the radio spectrum to which the receiver is tuned, is familiar to FM listeners because of regular announcements that the station is transmitting on such-and-such a number of megahertz, the same number that appears on the tuning dial. The term megahertz is made up of the international prefix *mega*, meaning one million, and *hertz*, the unit of frequency (one cycle per second) named after the German physicist Heinrich Hertz, who in 1888 first produced radio waves and demonstrated their major properties. Thus if the station announces that it is broadcasting on a frequency of 90.1 megahertz, it means that the radio waves emitted by the station are distinguished by the property that 90.1 million wave crests pass any given fixed point per second (there are 90.1 million cycles per second). The position of the FM band, which ranges from 88 to 108 megahertz, is shown in *Fig. 7* along with the TV bands. The AM band, not shown, stretches from 0.54 to 1.6 megahertz, and other services such as police radio, airport radar, ship-to-shore radiotelephone, radio aids to navigation, blind landing systems, military communications, communications with space probes, radio astronomy, international transmissions from one earth station to another, via earth satellite, walkie-talkies, etc., are sandwiched into other parts of the radio spectrum under strict national regulation and international agreement.

Any discussion of radio is bound to contain references to frequency, or its equivalent, wavelength. Conversion from *frequency* to *wavelength* (as expressed in centimeters) and vice versa is shown in *Fig. 7*. Also shown is the important microwave band, which extends nominally from a wavelength of one centimeter to a wavelength of 100 centimeters. The diagram does not extend far enough to show other electromagnetic waves, such as light, which are identical in nature to radio waves, differing only in their wavelength, which is about a million times shorter.

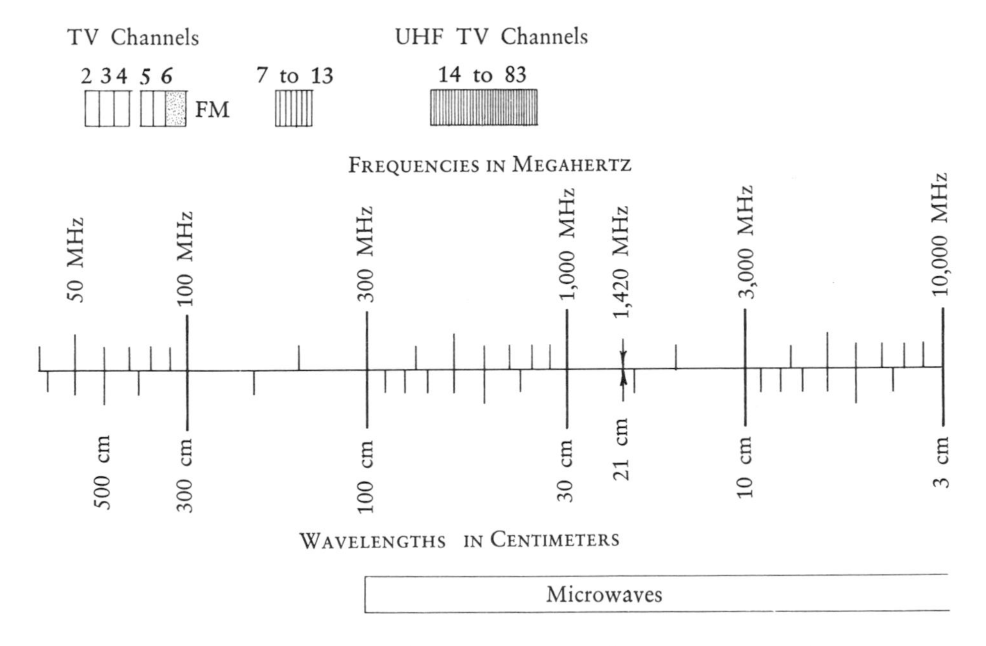

Fig 7. Radio wave frequencies. As it is equally common for a radio wave to be referred to by wavelength or frequency, this nomogram shows the conversion from one to the other. (The basic unit of frequency is the *hertz*, which used to be called the *cycle per second*. A megahertz, MHz, equals 1 million hertz.) The spectrum of electromagnetic waves continues to higher frequencies not shown here, including infrared, ultraviolet, visible light, X rays, and gamma rays.

Bandwidth

A TV set tuned to, let us say, Channel 4, having a frequency of 69 megahertz throughout the U.S., does not pick up adjacent channels at the same time—at least not if the set is in correct adjustment. This is because a TV set is intended to receive in a certain band only, the width of the band allocated to each channel being 6 megahertz. The band set aside for an AM station is only 1/1000 as wide as for TV. Bands much narrower still are used in space communications (some as little as a few hertz wide). Both narrow-band and wide-band receivers may be of importance to extraterrestrial communication.

Sensitivity

The third attribute of a radio receiver is its *sensitivity*, which refers to the faintness of a signal that the receiver can discern. Receivers of the highest sensitivity are desirable in order to reach far into space.

The most sensitive, widely tunable radio telescope receivers, combined with the largest fully steerable antennas, provide the optimal tool for radio communication with outer space. Such instruments are already in use in *radio astronomy, the study of extraterrestrial radio waves that originate naturally in thousands of celestial objects that have been discovered.*

Radio astronomy began in the 1890s as a theoretical notion accompanied by unsuccessful experiments, but the first observational discovery of *cosmic static*, or radio emission from the Milky Way, was made by accident in 1931 by Karl Jansky of Bell Telephone Laboratories in the course of a study of radio interference. Before the advent of radio astronomy, our picture of the universe was limited to a catalogue of *visible* objects that had been seen by or photographed with traditional telescopes. The variety of constituents that have revealed their presence through their radio emission has been surprising; their discovery has revolutionized our perspective on the universe.

The spiral structure of our galaxy, invisible because of clouds of galactic dust, has been discerned by radio telescopes tuned in on the faint emissions reaching us from the farthest outskirts of the galaxy on 1,420 megahertz (the natural emission frequency of hydrogen). Invisible galaxies on the fringes of the universe have made their presence known, and other galaxies, visible as points of blue light and emitting vast amounts of radio energy, have come to be known as quasars, but are not yet understood. The pulsars—stellar objects which, if listened to by radio telescope, tick like clocks of immense precision, and, if photographed with a sufficiently short exposure time, prove to be winking on and off—have contributed greatly to the understanding of supernovae, stars that explode to a brilliance rendering them visible in daylight. The explosions discharge matter into interstellar space where it becomes the raw material for the formation of new stars. But this new material is enriched with carbon and other vital elements, synthesized in the nuclear furnace at the center of a star, that will supply the necessities of life to a new generation of stars. The interstellar matter itself is a source of faint radio emission which has revealed the presence of ethyl alcohol and other molecules in space which had not been detected optically. Some of these molecules are sufficiently complex to be of interest in connection with the origin of life.

Many other astounding discoveries have been made and no doubt more remain—including the conjectural *black holes* thought to be the final relic of massive stars which, having spent their energy, collapse to zero dimensions, generating a gravitational attraction so strong that no matter, light, or other radiation can ever escape from them. Such an object could not shine by its own or by reflected light—hence the term "black hole." But it might be possible to deduce the presence of a black hole observationally by the effects of its gravitational field on nearby matter. Alternatively, perhaps the real black holes will prove to be not quite as featureless as the collapse of a symmetrical object would suggest.

Hitherto all the sources of radio waves that have been discovered have been of apparently natural origin. Something revealing the presence of intelligence might well be discovered by accident in the course of radio astronomical exploration just as occurred with the discovery of most of the principal classes of natural sources. In the absence of such luck we have been forced to form plans for deliberate discovery.

CHAPTER FIVE

PROJECT CYCLOPS

I think there is no question that we live in an inhabited universe that has life all over it.

George Wald

IN THE SUMMER of 1971 a remarkable band of 24 scientists and engineers, participants in the Stanford/NASA/Ames Summer Faculty Fellowship Program, assembled to study a system for detecting extraterrestrial life, under the joint direction of Dr. Bernard M. Oliver and Dr. John Billingham. Contributing to the technical power of the permanent group were 16 distinguished consultants from universities, government agencies, laboratories, and industry, who participated in a seminar series. The group investigated the various aspects involved in making radio contact with an extraterrestrial intelligence. Sections were responsible for studying the design and cost of large antenna structures, sensitive radio receivers, electrical transmission and control, and signal processing. One section worried about system design as a whole. The end product of the labors is a 243-page volume entitled *Project Cyclops*. By mining into this splendid study one can find in one convenient place the answers to many of the technical questions that continually come up in discussions about extraterrestrial communication.

Will Radio Waves Carry?

How far can one communicate by radio? Do radio waves become fainter with increasing distance just as light does, or are there obstacles in space that impede radio transmission? We know from observations of natural radio sources that there is no substantial impediment to the passage of radio waves through space. The only significant factor affecting their strength is the inverse-square law that applies to anything that spreads out in straight lines in three dimensions from its point of origin.

With this background we can calculate the strength of signal produced at a distance by a given transmitter, or by *natural* radio sources. This calculation is the basis of the design of radio or TV transmitters on earth, but in fact it is even simpler in space because the effects of the earth's terrain, its curvature, and the influence of its atmosphere do not introduce complications.

If a replica of the 250-foot radio telescope at Jodrell Bank in England, complete with 100-kilowatt radio transmitter beeping once a second, were to exist on some distant planet and were to beam a microwave signal to the 328-foot diameter radio telescope in West Germany with its high-quality maser (amplifier) receiver connected, how far away could the transmitter be and still be able to deliver a detectable signal? (For this calculation I have selected the largest fully steerable radio telescope antennas now in actual existence and have assumed the best current receiver technology.) The answer is that the transmitter can be approximately *400 light-years* away from the receiver and still deliver detectable beeps. This is a tremendous distance, encompassing approximately 100 thousand stars and their planets. While it is only a fraction of the size of our galaxy, it still amply covers the volume of space within which man might hope to establish communication. After all, a message received from that distance would take 400 years to arrive and the round-trip time for a reply would be a minimum of 800 years. Radio communication is thus perfectly possible as a mode of communication in space.

Many other means of communication are discussed quantitatively in the Cyclops report, including waves of other frequencies such as light and infrared radiation, which can now be generated by lasers producing very powerful beams. Such engineering discussion requires careful attention to many technical aspects: the sensitivity of present receivers, the size of telescope structures, the interference generated by the universe itself, and the allowable error rates associated with the various sophisticated electronic signal-processing systems. Radio is found to be superior to these alternatives, particularly in the microwave range of wavelengths from about 10 to 30 centimeters. This is precisely the frequency range in which many major radio astronomy observatories conduct research. Thus, the familiar big radio telescope dishes give us an idea of the nature of the tools that could be used for interstellar communication.

The Search Phase

Proceeding from these basic considerations the Cyclops team turned its attention to design questions for a major radio antenna installation within the capacity of human society to build in the near future and offering hope of effecting contact with intelligent extraterrestrial life.

New types of questions now arise. In our elementary calculation of the range of communication possible with radio transmission we assumed that the equipment at both transmitting and receiving ends of the link was known to both parties. Thus, to be precise, those at the transmitting end would know:

a) What frequency to use
b) Where to point
c) When to transmit
d) How the message is encoded (a term embracing technical details such as bandwidth and form of modulation—e.g., AM, FM, or other)
e) What language to use

Those at the receiving end would know these same factors. We found no difficulty covering immense distances under these *favorable* conditions, which arise under what we describe as the *communicative phase.*

But, before the information listed becomes known to both the parties—as is the case in searching for intelligent extraterrestrial life—the situation is very different, and the problems are most difficult. We refer to this as the *search phase.* The Cyclops system is designed for search but, if and when contact is made, the proposed installation would be far more powerful than needed for communication and could be diverted to study of interstellar molecules and other branches of radio astronomy.

Let us examine the choice of frequency and simplify the problem of where to point the telescope by first considering the case where the other party either knows or has guessed that there is a civilization on earth. Telescopic observation by an outsider over a period of time might conceivably reveal changes as man transforms his planet, or we might simply be on a catalog of likely prospects. But for whatever reason, the outside intelligence is irradiating the earth with a permanent radio beacon to inform us of their presence. If we knew the frequency, our chances of detecting such a beam would be favorable. However, as we don't know, we would have to search the radio spectrum. It is thought that approximately 2 billion channels might have to be examined in the microwave range from 10 to 30 centimeters wavelength mentioned above. While this is a good deal more than the number of channels we are accustomed to in domestic broadcasting in the TV, FM, and AM bands, it is not just a matter of twirling the dial; much more is involved in finding a contact message than in finding a station on the radio. We would need to listen to all 2 billion possible channels in each of the large number of directions where there are suitable stars. No doubt each sky direction would have to be returned to at another time.

Thus, it is of prime importance to know the frequency of transmission. Much discussion has taken place about this matter. In 1959 Giuseppe Cocconi and Philip Morrison developed the following argument:

> The other party realizes that it would be risky to transmit on any random wavelength and will therefore choose a wavelength that has some *universal* significance. Hydrogen is the most abundant constituent of the universe, as all inhabitants will know regardless of where in the universe they may be, and the hydrogen atom has the ability to generate microwave radiation spontaneously with a frequency of 1,420,405,750 hertz (the corresponding wavelength of approximately 21 centimeters is in the middle of the optimum low-noise band). This frequency therefore has a *universal uniqueness*, not set by anthropocentric considerations, that fits it as the outstanding choice for prospective communicators who have not had an opportunity to agree on a frequency. (*Nature*, vol. 184, p. 844. Italics added.)

This ingenious and beautiful thought is so striking that it was responsible, I believe, for the resurgence of interest in extraterrestrial life that has taken place in the years since then. Some of the newly stimulated attention has focused on reevaluation of their argument on technical grounds and has led scientists to advance reasons for going to twice, or one half of, the hydrogen frequency. Oliver believes that a contact signal may be found not precisely at the hydrogen frequency but in a band stretching from that frequency to certain frequencies just below 1,700 megahertz associated with OH (hydroxyl) molecules. Since the chemical combination of H and OH results in water, H_2O, he quaintly refers to this low-noise part of the spectrum as the waterhole and waxes lyrical over the thought of two civilizations coming together to meet at the waterhole.

For practically all human societies, be they farmers, herders, hunters, or nomadic gatherers, the waterhole has deep meaning with connotations of survival, but whether such feelings exist in alien civilizations or would influence their choice of frequency remains to be seen. It cannot be denied, however, that our first encounter with another intelligence has more to it than sheer technical aspects. The whole thought of encountering the unknown has a strong element of poetry.

The Cyclops group does not subscribe to the view that these self-advertising frequencies can be counted upon. Instead, they propose a receiver system designed to cover the wavelength range from 10 to 60 centimeters (500 to 3,000 megahertz). Ingenious detection schemes are then proposed that can handle a 100 megahertz band containing 100 million channels *simultaneously*! Then, listening for a time of the order of 1,000 seconds (16⅔ minutes) to each target star, scientists can work through a long list of possible stars, covering ranges out to hundreds of light-years in about a decade.

The Cyclops System

The whole proposed Cyclops system would comprise 1,000 or more large radio telescope dishes spread over an area up to 10 miles across, at a projected cost of $6-10 billion. Although the construction cost would be about one-tenth of the U.S. national budget it would be spread over 10 or 15 years.

Existing star catalogs, which list hundreds of thousands of stars and represent the painstaking accumulation of centuries of observation, will not suffice for the Cyclops program. The list of target stars needed would have to supply both the spectral type (such as F, G, or K) and the distance, to make possible an orderly search procedure, beginning with the most favorable combination of nearness and spectral type. Therefore, to supplement and extend current astronomical knowledge, the Cyclops report proposes the construction of a computer-controlled observatory, described as follows:

> The system we visualize consists of a battery of telescopes each supplied with automatic positioning drives, a computer, tape reader, and a copy of part of the target list. Each telescope would go down its portion of the target list looking successively at the stars visible at that time of year. The star's spectrum under low or medium dispersion would be imaged onto the target of a long time-constant (cooled) vidicon for the time needed to obtain a good record of the spectrum. The exposure time could be controlled by letting part of the light of the un-

> dispersed star image fall on a photomultiplier and integrating the output current.
>
> When the spectrum recording was completed it would then be scanned by the vidicon and stored in the computer memory. There it would be cross-correlated with standard spectra taken by the same telescope system from stars of known spectral type. The spectrum showing the highest cross-correlation would then be used to classify the star being examined. If the star were indeed a main sequence F, G, or K star, the distance would be computed from the exposure time required and the star would be entered on the new master tape listing. If not, the star would be rejected.

In other words, the spectrum of each star would be received on a television camera for a length of time appropriate to the faintness of the star, but instead of being displayed on a TV tube, the spectrum would be fed into a computer. The computer would match the spectrum with samples of different spectral types stored in its memory. The computer would reject stars not of spectral types F, G, or K, calculate the distance to the remainder, and list them on a master tape.

The report continues:

> If the refining of the list is to keep pace with the search and if the average exposure time is equal to the search time, three telescopes would be needed, even if there were no false entries on the original list, because Cyclops can work all day and all night, while the telescopes can only work on clear nights. If, in addition, the refining process rejects half the original entries, then the telescopes must examine twice as many stars as Cyclops, so six telescopes would be needed.

The startling spectacle of six telescopes of major proportions (more than the whole United States now has) busily opening and closing their own domes, pointing themselves about the sky automatically and tape recording their findings conveys the scale of this ambitious project. The recommendations of this substantial study provide food for thought and discussion which, in view of the total cost, will no doubt go on for a number of years. Strongly influencing the outcome of the debate over Project Cyclops will be the probability of its success. We are left with the simple question: What are the real chances of establishing radio contact?

PARIS

CHAPTER SIX

MAKING RADIO CONTACT

Perhaps the safest thing to do at the outset, if technology permits, is to send music . . . I would vote for Bach, all of Bach, streamed out into space, over and over again. We would be bragging, of course, but it is surely excusable for us to put the best possible face on at the beginning of such an acquaintance.

Lewis Thomas in *Lives of a Cell*

IN PROJECT CYCLOPS you have seen an extensive study by a group of competent engineers and scientists, resulting in a proposed system for establishing contact with an external civilization by radio. Communication over interstellar distances by radio *is* possible, and microwave radio is the best part of the electromagnetic spectrum for the purpose. But in the *search* (or pre-contact) phase, as distinct from the *communicative* phase that would develop *after* the initial contact had been completed, *substantial difficulties are encountered, one of which is the choice of radio frequency*. The Cyclops system proposes to overcome this difficulty by scanning in frequency during the time allocated for observing each particular star. We may picture this as equivalent to slowly turning the tuning knob until the microwave band is covered. Alternatively, a wideband receiver may be used to receive simultaneously all frequencies in a certain band. After the allocated time, the receiver would jump to the adjacent band, and so on. The Cyclops group worked out a very sensitive and sophisticated electronic system for finding a faint signal.

I believe this scheme would work well if there were a community within 30 light-years of the earth (we will call this Case I) *and* if this community were beaming a powerful signal steadily in our direction. As there are only around 50 habitable stars within that range, a determined search might succeed in making contact within a year. Such a success would undoubtedly cause a great flurry of excitement, but there would be ample time to deliberate the next course of action since up to 30 years would elapse before "they" could receive our response and, even assuming they picked up this response and reacted promptly, a further 30 years would pass before we heard their "Hello." Needless to say after this 60-year lapse few if any of the original participants would be around to share in this further excitement, and those who were would be beyond retirement age.

Great perseverance would be required on the part of the extraterrestrials at their transmitting end. Would they have maintained their radio beacon throughout the million-plus years that intelligent life has existed on earth, waiting for the days when Maxwell would predict radio in 1873, when Hertz would make radio waves in 1888, when Marconi would communicate by radio across the Atlantic in 1901, up to the time 19?? when the Cyclops system would be turned on and pointed at their planet? Or did they simply monitor our world at intervals until, when the age of broadcasting began, they were able to pick up *faint signs* of our beginning activity in the radio spectrum?

Leakage Radiation: A Self-Advertisement

What would these faint signs be? A single 50-kilowatt TV transmitter spills much of its energy over the horizon. The thin sheet of leakage radiation (see *Fig. 6*) gyrates around the sky, as the earth rotates on its axis, illuminating a distant planet for only about 20 minutes, twice a day. The Cyclops report shows that this unintentional leakage radiation could be detected 30 light-years away with a high-quality radio receiver tuned to the right frequency, if the receiver were connected to an antenna 3 kilometers in diameter pointed in our direction. There would be problems for the extraterrestrials to get onto *the right frequency at the right time.* Nevertheless, the fact that our leakage radiation is strong enough to be detected makes it conceivable that it would be.

Some writers conjure up a graphic picture of a great onion that has been expanding around the earth at the velocity of light since 1901,

whose successive layers contain news of world events, old radio serials ("The Lone Ranger"), sports reports, the music of the Beatles, and so on, that could be tapped by some interstellar agency. They might certainly detect our presence, but the programming would mostly go over their heads because rotation of the earth would limit their coverage of any one station to the approximate 20 minutes, twice daily. After this initial detection took place, perhaps their beacon would be turned on. What frequency might they choose for their beacon?

Where Is Their Beacon?

Possibly they would choose to tune their beacon somewhere in one of our TV bands. Therefore, we should all be alert for the first message, which may show up on the TV set of any one of us. During regular program hours we might interpret extraterrestrial signals merely as troublesome interference; conditions would be more favorable for reception late at night after the local stations have gone off the air. Occasionally one may catch glimpses of programs on vacant channels, usually coming from another station. Such reception of a remote station can occur due to unusual atmospheric conditions, or as a reflection from transient trails of meteors plunging through the upper atmosphere. In view of these exceptional possibilities it would be helpful to know what to expect in the way of an extraterrestrial message as distinct from a terrestrial program.

What Will Their Message Say?

In 1941 Sir James Jeans reasoned that we could attract the attention of the Martians "if any such there be" by shining a group of searchlights toward Mars and emitting flashes to represent a sequence of numbers such as 3, 5, 7, 11, 13, 17, 19, 23 . . . , the prime numbers. Subsequently, other authors have suggested that extraterrestrials might use this same type of message to contact us. Personally, I think it would be rather anticlimactic for designers of some high-power radio transmitter in space to use their program time trying to prove to me that they could also *count*! At the least I would expect a little poetry or art. In any event, let's give them credit for enough imagination to put on a program that would rivet our attention.

Another thought regarding the message's content is based on the supposition that the beacon will have to remain turned on for a very long time before any acknowledgment is received. A dilemma faces our extraterrestrials. A long story runs the risk that we tune in near the end. A short one repeated again and again bores us to tears for decades while we try to acknowledge. This dilemma has led to a further idea: messages might be nested within messages—short items, frequently repeated, sandwiched between episodes of a longer story repeated less frequently, all of which is contained within an even longer communication, and so on. Thus, no matter when we tuned in there would always be enough variety and recapitulation to keep our attention.

One in a Million

The weaker links in this scenario for making radio contact become increasingly troublesome as we proceed to greater depths of space, say in the range 30 to 300 light-years (which we will call Case II), where we are necessarily dealing with fainter signals. At our end is the greater difficulty in discerning signals arriving from afar; at their end is the increased difficulty of knowing that the time has come to transmit their message to our planet. A geometrical problem also presents itself. At our end, to search to a depth of 100 light-years, we must scan approximately 1,000 likely stars; at their end, an equally great number of possibilities lie within the sphere of a 100-light-year radius surrounding their home base. There is one chance in a thousand that they will be transmitting in our direction, and one chance in a thousand that we will have chosen to look in their direction at precisely the time when their transmission arrives in our neighborhood. The combined chance of this occurrence is one in a million.

As an indication of the amount of time likely to be allocated per star we may take the illustrative value of 1,000 seconds adopted in the Cyclops report. Obviously, if the beacon was beamed and was working its way through a list of prospects while we were working our way through our list of prospects, the chances of our locking horns in a given time-slot of 1,000 seconds would be extraordinarily slim, requiring about 30 years of perseverance at both ends to achieve a hit. This handicap is aggravated by increasing distance and ultimately becomes overwhelming. In addition to this geometrical problem, one party has to choose a frequency and the other has to include that frequency in the range of frequencies that it guesses.

An interesting paradox is that while much radio-frequency power is escaping from the earth, it is contrary to good engineering practice to so waste power. The trend is to supersede broadcasting by improved point-to-point methods of communication that would exclude the radio-frequency power leakage that might enable other beings to know we are here. Cable TV, submarine telephone cables, and earth stations interconnected by microwave beams via relay satellites all furnish examples of this trend. The earth could cease to be a self-advertising source of radio leakage. Still the frequency bands vacated might be occupied immediately by radio services such as aircraft collision avoidance radar or automobile radiotelephones which, by their nature, cannot use point-to-point communication techniques.

Realizing that the beacon people may not know where we are, the Cyclops report studies the case of a beacon radiating about 1,000 megawatts divided equally in all directions, on a steady, permanent basis. Naturally, this is costly to the extraterrestrials in energy and would result in a fainter signal at our end than if they beamed the same total energy at us only, but we could detect their beeps from 500 light-years away with the Cyclops system, which is capable of reaching well beyond our Case II of 30 to 300 light-years. Use of an omnidirectional beacon appears to reduce the geometrical dilemma encountered earlier. Yet, does it really? *At a time not known to them in advance, our reply acknowledging receipt of their initial message will one day fall on their planet, and they must be ready to receive it.* Not only do they not know when our reply will come, they do not know from which direction it will come, as their beacon has been transmitting with equal intensities in all directions. They must therefore have a great receiving antenna capable of being pointed in various directions and, in accordance with some schedule, they must switch their beam around the sky from star to star using a list that they have prepared. They must dwell on our direction long enough to detect our faint answering call, if it is present, and then move on to the next candidate and so on down the list until the whole circuit of the sky is completed. At the end of this sequence of observations they must repeat their scanning of the sky and devote more time in our direction again before moving on. It would be too bad if our reply fell into the time slots devoted to other stars. But at least they can count on our understanding this problem, and it would be a wise move on their part to permanently encode the length of their repetition interval into their beacon transmission in some way so that we would know how long our answer must be sustained before there could be a chance of its being heard.

Political Headaches

I am troubled that we might not be able to fund the necessary return transmission for the necessary length of time. How can a national government or international agency justify continued annual expenditures on a project whose success cannot be known for 60 to 600 years into the future (bearing in mind that in this continued discussion of Case II we are considering ranges of 30 to 300 light-years). I also wonder about the *presumed* political stability at the other end; can an omnidirectional transmitter blindly radiating 1,000 megawatts in no particular direction

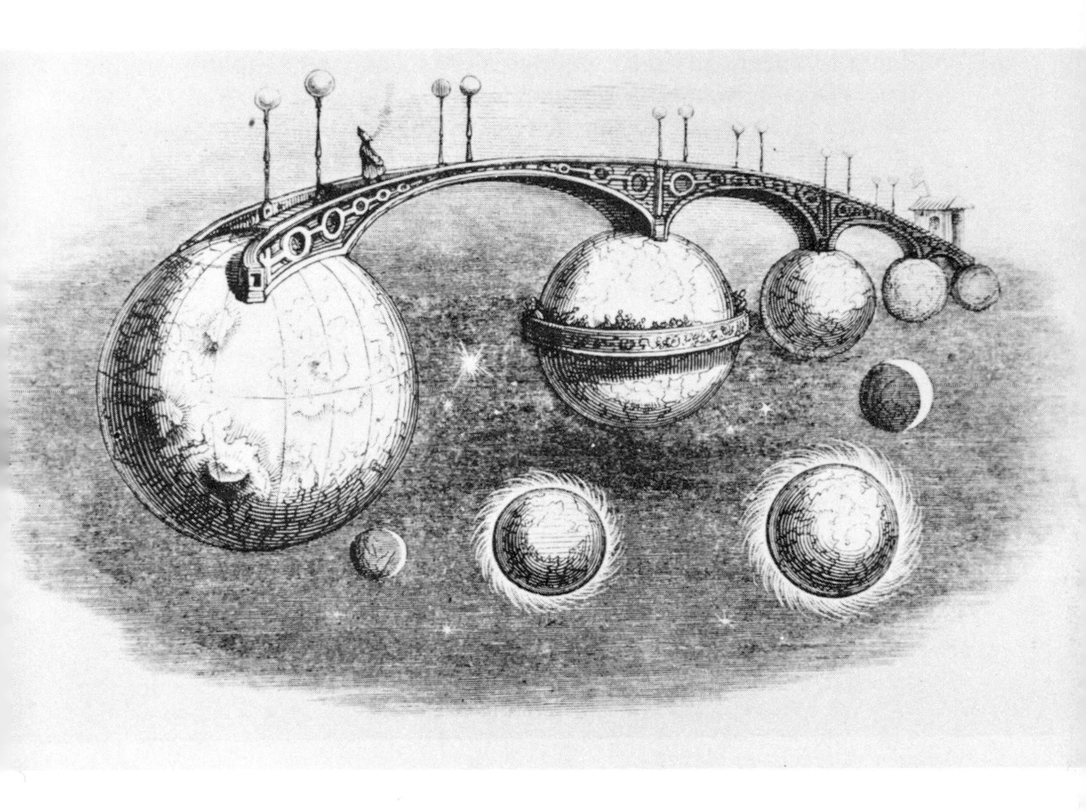

be justified and sustained for centuries on end? A weak link is the limited fraction of time that they will devote to listening in our direction. In addition, as we do not know their antenna size, we must be sure to send enough power. Consider another problem that might arise: Imagine our House Appropriations Committee reducing the proposed expenditure for the project and the Senate (realizing success is thus jeopardized) deleting the item entirely for reconsideration the following year. Even if these problems are surmounted and our great transmitter does get funded, what will a review board conclude 10 years later when new faces occupy the Office of Management and Budget as well as the various congressional committees? Those in favor of continuing will have no progress to report and can expect none for *50 years or more*! There is a further possible problem of morale in the National Center for Exobiology, whose responsibility it will be to conduct the transmission, because the younger scientists may wish to divert some of the budget to new and exciting questions of solar system exploration, prebiology, and even life forms that may have been discovered on some satellite of Jupiter. They may resent the stodgy attitude of the conservative senior staff, who spend large sums on electric power bills for the transmitter the same way year after year after year! These uncomfortable political-economic omens will be seen to have their root cause in the long round-trip time, and they become very serious when the distance separating us from the presumed omnidirectional beacon exceeds, let us say, 30 light-years. For distances greater than 30, say out to 300 light-years, the scheme does not sound like one that society as we know it historically is geared to support.

Among the questions aired in congressional committees which could jeopardize a prompt response might be the following:

Senator X: Now, if I understand correctly, you believe that this foreign civilization has scheduled a quarter hour to look in our direction on a certain date a hundred years from now. What happens if they have a technical fault at that time?

Dr. Y: On a matter of such importance they would surely have a backup system but if, for any reason, they missed our transmission they would pick it up in their following fiscal year.

Senator X: So we simply have to fund the transmission for one or two of their fiscal years.

Dr. Y: It would be better for us to transmit for 200 of our years, otherwise all the time would be wasted while we are waiting for their reply.

Senator X: When their reply comes, could it happen that the United States would be on the side of the earth that is facing in the other direction?

Dr. Y: Yes, that is why the International Astronomical Union suggests cooperation with China and the Soviet Union, each to receive one-third of the message. Of course, they will have to install extensive receiving antenna systems similar to ours, but we have almost 200 years in which to negotiate the details of Project Triclops, as we call it. If the plan falls through, the United States can always go it alone, putting the whole system in orbit.

Senator X: Is there any technical reason why other countries should not duplicate the whole system in orbit, so that all could receive the whole message?

Dr. Y: No. As a matter of fact, that is the French proposal.

Senator X: The Federal Communications Commission is not prepared to clear the extraterrestrial band for exclusive use now and refuses to make rules that purport to be binding two centuries into the future. What is your response to that?

Dr. Y: If, as we expect, the message arrives with a 100 megahertz bandwidth centered on the waterhole frequency, with high intensity and high information rate, it will automatically, though regrettably, force airport radars and other services using that band off the air. But we would prefer not to design the new Triclops systems for that band because, if the message is weak, interference by terrestrial users of the band may be overwhelming. Experience with attempted worldwide protection of bands for radio astronomy engenders no degree of conviction that the extraterrestrial band could be cleared. However, there is a suitable military band in the United States that was formerly allocated to the now obsolete Nike coastal defense radars. We hope to induce the beacon people to use that band by placing our reply in it, should these hearings result in approval of our budget.

Senator X: Can any nation send a reply?

Dr. Y: At the recent General Assembly of the International Scientific Radio Union, representatives of 14 countries confirmed that the beacon had been received on the precise frequency originally announced by my colleague Dr. Z and several governments are now considering action.

Senator X: It might be wise to avoid hasty unilateral action at this point in time.

Would this be so very unlikely?

CHAPTER SEVEN

INTELLIGENT NEIGHBORS: HOW FAR, HOW MANY?

I make a sharp distinction between intelligence and technology. It is easy to imagine a highly intelligent society with no particular interest in technology.

Freeman J. Dyson

WHEN GIUSEPPE COCCONI and Philip Morrison suggested that a technical civilization in our neighborhood was beaming a message at us (*Nature*, 1959), a flurry of intellectual activity was stirred up that has not subsided to this day. Although scientific speculation about life in space was by no means absent before that time, there was something about the new paper that made the prospect of contact seem more imminent. Perhaps it was the persuasive argument that 1,420 megahertz, the frequency of hydrogen, is the frequency to tune in on. Or perhaps the feeling of imminence had been present before, during the era of telescopic mapping of Mars, and a later generation of scientists was now ready, waiting to be sparked.

How Far Away?

Cocconi and Morrison did not place much emphasis on a certain vital parameter, namely the *distance* to the nearest superior technical community, but one can deduce that they had in mind that the nearest such community would be from 5 to 50 light-years away. If we accept the notion that a community *technologically* more advanced than ours would indeed be interested in attracting our attention, it seems to me that the most suitable means of their doing so might depend on the distance d from the earth to the nearest *superior community. We define a superior community as one surpassing our current level of technological ability.* I believe there are three distinguishable possibilities of such life within our own galaxy, as I reported in a *Nature* article in 1960. The distinction between these galactic cases will be discussed below. If, however, the nearest superior community is *not* in our galaxy at all, but outside it, we have Case IV. Finally, Case V covers the remaining possibility, namely, that no community superior to us exists, anywhere!

Table 1: Possibilities for the Incidence of Superior Galactic Life

	Superior Technical Life Within Our Galaxy			Special Cases	
Case	I	II	III	IV	V
Degree of abundance of life superior to us	Abundant	Sparse	Rare	We are alone in the galaxy	We are alone in the universe
Distance to nearest superior community, d	Less than 30 light-years	From 30 to 300 light-years	From 300 light-years to edge of galaxy	Greater than 1 million light-years	∞

This table sets forth the basis for the discussion in this and subsequent chapters. Clearly the distance d to the nearest superior community is a well-defined quantity, but we do not at present know how many light-years this distance is.

Technical Civilizations: How Long-Lived?

In the course of surveying the range of possible distances, I realized that there is a critical connection between the distance d and the average *lifetime* of a community, *as measured from the time when it passes through our present level of technology to the time when it either expires or decays.* Of course, it might be difficult to determine at precisely what moments these initial and terminal events occur, since on another planet technical achievements might take place in a different sequence. Perhaps rockets could have been launched into orbit before radio was discovered, or nuclear energy might have been discovered before airplanes were invented. There might be brief periods when it would be difficult to say whether a given community is ahead of another or not. But technology moves so fast that the time of arrival at technological superiority must be sharply defined in the time scale of the whole of history.

Even so, it is amusing to contemplate the sequence of events on a planet where the circumstances do not exist which led man to the discovery of electricity and magnetism and then to electrical machines and radio communication. Electricity has always been part of man's experience (through lightning or electrostatic phenomena such as the sparks seen when a cat is stroked at night), but it could be absent on a planet with a damp, salty atmosphere. Likewise, on a planet without a magnetic field, the inhabitants would be deprived of the hint furnished by lodestones which, suspended on threads, were the first compasses. On such a planet clockwork and fine mechanism might develop to surprising levels before electricity was hit upon. (Would it be possible to make a transcontinental, nonelectric telegraph line, starting from the principle employed in the speaking tube that a ship's captain uses to deliver orders to the engine room? No doubt amplifiers would be needed to carry a message over thousands of miles: Could they derive their power from a wound-up spring rather than from electric batteries?)

The existence of a superior community may terminate in various ways. But whether the technological capability declines gradually or stops suddenly, the lifetime of the superior community *ends* when it ceases to perform technologically at a level above ours at present.

Not all superior communities will have the same longevity. Some communities may die young. The possibility of premature self-annihilation on earth has been a serious political concern since the advent of nuclear explosions. (The explosive power of nuclear bombs is now equivalent to a ball of dynamite 7 feet in diameter for each man, woman, and child on earth.) Some communities may survive for millions of years before succumbing to an inexorable natural phenomenon such as the running down of the energy source in their life-supporting star. Still others, it is possible to imagine, may learn how to circumvent this fate, perhaps by colonizing a younger star. There may even be durable communities that emerged at the earliest possible moment after our galaxy began (some 10 billion years ago) and have achieved quasi-permanence. For all we know, many superior communities may survive indefinitely. But I will pursue the consequences of assuming that it takes 5 billion years *on the average* for likely stars to evolve a superior community, which then endures for an average lifetime L measured in years. (The figure of 5 billion years is the age of the solar system, and the term *likely star* is a brief way of referring to those stars of our galaxy, taken to be 10 billion, that may have planets in stable orbits.)

The Vital Connection

The connection between the distance d of the nearest superior community and the average lifetime L (referred to above) comes about in the following way: Take the case where $d = 100$ light-years. From the known distribution of stars in our galaxy one can construct a graph, shown in *Fig. 8*, and read from the point marked B that there are *1,000 likely stars* within *100 light-years* from the earth. The 100-light-year example that we have selected for specific discussion falls under Case II of *Table 1* where the degree of abundance of life is characterized as *sparse*. We can now add precision to the characterization by noting that in this example there is one superior community per 1,000 likely stars.

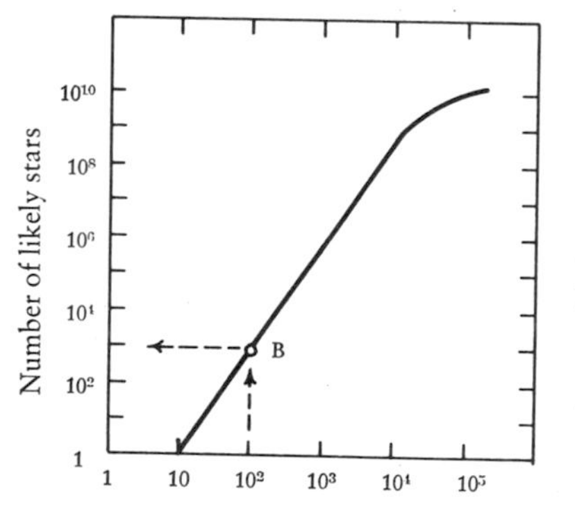

Fig. 8: The number of *likely stars* within an earth-centered sphere of given radius.

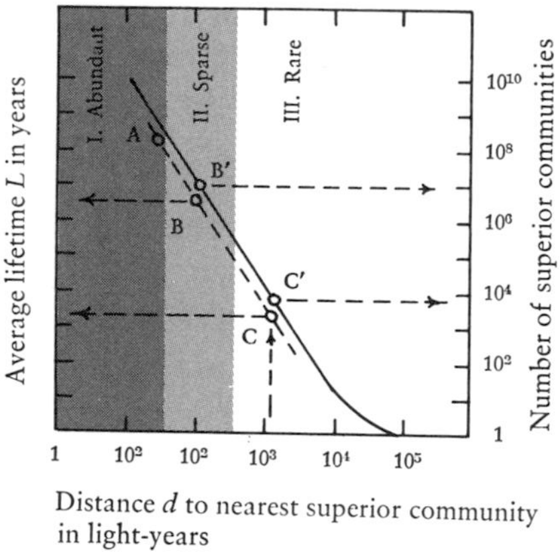

Distance d to nearest superior community in light-years

Fig. 9: The total number of superior communities in the galaxy and the average lifetime L as they depend on the distance d to the nearest superior community.

Recalling that there are 10 billion likely stars in the galaxy, we deduce further that an occurrence of one superior community per 1,000 likely stars gives us a total of 10 million superior communities for the whole galaxy. With 10 million such communities distributed about the galaxy, an average lifetime of 5 million years (a thousandth of the 5 billion year evolution time) is required to maintain an equilibrium between birth and death rates—what could be referred to as galactic ZPG (zero population growth). Thus $L = 5$ million years when $d = 100$ light-years. Point B on *Fig. 9* represents this calculation. We can ascertain the connection between lifetime and distance for any example falling into Cases I, II, or III from the same graph. This graph will be called on when we consider the implications of the different cases for interstellar communication. A second curve presented in *Fig. 9* enables us to determine the total number of superior communities, whatever the distance to the nearest one; point B' (10 million communities) refers to $d = 100$.

I have used 5 billion years as a representative time for a superior community to evolve because it applies to our earth, the only intelligent living community we have information about. Perhaps the evolutionary time averages as little as 2 billion years for the galaxy as a whole; if so, an average lifetime of 5 million years would bring our neighbors closer to us; for example, the 100 light-years would be reduced to 75 light-years. Other corrections might be required, for example, if actual birth and death rates of different star types were used to supersede the calculations based on equilibrium.

Threshold Density: Life Is Rare

One of the things we discover from *Fig. 9* is that when the distance d is 1,250 light-years, the average lifetime L is equal to 2,500 years (see point C), which is precisely the round-trip travel time of radio waves to that distance and back. Consequently, if at the moment of attaining our present level of technological capability a community at a distance of 1,250 light-years from us emitted a signal that we received and replied to *immediately* on receipt, our reply would arrive *just as they went under*!

The critical conclusion is that *a certain threshold density of superior communities has to be reached if the members are to be able to converse; otherwise their average lifetime will generally not be sufficiently long for a message to traverse the vast depth of space to the nearest neighbor and return.* This condition sets in when the distance is about 1,250 light-years to the nearest superior community. Although *Fig. 9* (at point C') implies that at this distance of 1,250 light-years there would be as many as 5,000 superior communities in the whole galaxy, that is only 1 in 2 million of the 10 *billion* likely stars. In this case, we could say (as in *Table 1*) that superior life is rare and therefore the average community is unable to enter into two-way communication. We must, however, allow for some contacts arising from accidental proximity, and we cannot eliminate the possibility of a distant race emitting a swan song which may or may not be heard—and to which it expects no reply.

Abundant Life

Shifting attention now to Case I, Abundant Life, we find from *Fig.* 9 (point *A*) that if the nearest superior community is only 30 light-years away, an average lifetime *L* of 200 million years is implied, and there would be 300 million communities superior to us in our galaxy. In fact, one in 25 of all likely stars would have such a community. Of the remaining 24, some might have some form of organized life evolving toward higher forms, and the rest would be in some stage of geological, chemical, or prebiological evolution. Such abundant life seems unlikely because it does not seem to make enough allowance for the dead ends, short circuits, and other perils besetting the upward path toward intelligent life.

If Case I, Abundant Life (30 light-years or less to the nearest superior neighbor), does turn out to be the actual situation, I believe that either we *or* they can attract the other's attention by pointing a radio beam and blazing away on some frequency. Strategies such as those devised under Project Cyclops should succeed promptly, since so few stars will need to be scrutinized and the financial outlay can be modest. After all, if Case I is true, the sought-for neighbor will be among the *first* 25 on our list of prospects and so close as not to require antennas any larger than are already available in radio astronomy observatories and at NASA communications stations.

Sparse Life

Case II, Sparse Life (30 to 300 light-years to the nearest superior neighbor), is potentially the most important. The prospects for establishing communication if Case II applies are good near the 30 light-year end of the range. Difficulties then set in, including the geometrical problem previously mentioned, and they become more grave as the distance increases. Finally as the 300-light-year end is approached, any dialogue would be limited to the lifetime remaining after contact was made.

Durable Communities?

The concept of average lifetime of a superior community and the deduction that it must be connected with the number and spacing of such communities have been picked up and discussed by prominent scientists such as Sebastian von Hoerner, I.S. Shklovsky, and Carl Sagan—all of whom have also recognized that the subject subdivides, according to the closeness of spacing, into categories. I suggested in my *Nature* article that even though technology might be rare in the galaxy (Case III above), there may be some communities that have achieved durability, even quasi-permanence, perhaps by gaining control of the cir-

cumstances that lead to short average lifetimes. Aided by accidental proximity due to random spacing, some of these could be in contact. Presumably such an ancient association would be very able indeed technically.

A favorable location for such chains of communication would be toward the galactic center, where star density is much higher and the distance between neighbors is therefore much less than the average distance in the galaxy as a whole. This exceptional formation of chains, despite the rarity of technology, could lead to longer lifetimes because the members could help one another. Von Hoerner is pessimistic about the occurrence of galactic chains of communication even though the concept that contact between civilizations may help their longevity is one that he develops mathematically and calls *feedback effect*. Shklovsky and Sagan, on the other hand, calmly contemplate longevities of superior communities equal to the longevity of a star and subscribe to the notion that *communicative feedback* could be mutually beneficial in extending longevity.

When we have colonized interplanetary space—which could be early in the 21st century, according to Princeton physicist Gerard K. O'Neill's timetable—we will have concomitantly achieved independence of the terrestrial catastrophes that lie ahead. Survival of the fittest, on a time-scale of geological upheaval, may mean that communities over a certain age will be those that have succeeded in colonizing interplanetary space. To survive the decline or explosion of one's star (the communities 100 million years old or so will be those that have evolved to cope with this eventuality) it is merely necessary to move into interstellar space or into the environs of another star, particularly a long-lived star such as a slow-burning red dwarf (M-type star).

How to extrapolate galactic "survival of the fittest" to the whole universe is beyond me. By the time life and the galaxy are symbiotic, the galaxy will have been transformed into a living entity. Physics will be different. One can only suppose that the upward path of life would not then stop at the limits of our galaxy.

The Trail of Ideas: A Personal Account. *Cocconi and Morrison triggered a sequence of events that some scientist may someday wish to chronicle. A number of authors produced papers on the possibilities of extraterrestrial life and interstellar communication, and although their thoughts at first glance may seem unconnected, interesting relationships are present. I recall discussing the Cocconi-Morrison article with Stanford colleagues in 1959 and developing a number of ideas in quick succession in the course of conversation and seminars. Each person has his own enthusiasms, so at the time it did not strike me as strange that I was paying more attention to the possibilities of communicating with extraterrestrial life than my friends. Looking back, however, I realize the odd coincidence that in 1954-55, when I was lecturing* *on radio astronomy at Berkeley, two people were there who also became key figures in the new developments.*

Otto Struve, the distinguished Astronomy Department chairman who had invited me to come from Australia to Berkeley, had known since the 1930s that the slow rotation of stars such as the sun might be connected with possession of planets, and no doubt it was his influence that caused associate Su-Shu Huang to think about the habitable zones around stars. The appearance of Huang's series of papers on this subject in 1959-60 was independent of Cocconi and Morrison. I believe that my association with Otto Struve and Su-Shu Huang, both of whom I came to know quite well, must have prepared my mind for the subject of life in space.

In March 1960, Struve was director of the National Radio Astronomy Observatory at Green Bank, West Virginia, when I spoke there and gave him the manuscript of my article that appeared in Nature, *May 1960. At that time Frank Drake (who, as a Cornell undergraduate, had heard Struve's conjecture about planets) was preparing for Project Ozma, the first search by radio telescope for extraterrestrial life. Sebastian von Hoerner, involved in advanced radio telescope design, was no doubt turning extraterrestrial ideas over in his mind, because, by the time of the General Assembly of the International Astronomical Union in Berkeley in the summer of 1961, he gave me a manuscript of an important article on the subject that appeared in* Science, *December 1961.*

I am less familiar with the Soviet threads in the skein. In January 1961 while visiting Soviet radio astronomers, I spoke with I.S. Shklovsky's group at the Sternberg Astronomical Institute about extraterrestrial life. In due course Shklovsky devoted a chapter of his 1962 book to this material, which was largely carried over into the translation edited by Carl Sagan. Perhaps Struve, who grew up in Russia, and Shklovsky were influenced by Tsiolkovsky, whose prescient 19th century writings on rocketry, space exploration, and life on other worlds were ahead of his time. Tsiolkovsky corresponded with A.N. Tsvetikov, who was at Stanford in the 1960s in biophysics, where Joshua Lederberg was working on detection of life on Mars, Carl Sagan collaborated with Lederberg during 1962 while at the Stanford Medical School.

Enough has been said to show that the threads of ideas weave an intricate and interesting pattern, much of which the single participant never sees.

"Miss! Oh, Miss! For God's sake, stop!"

CHAPTER EIGHT

INTERSTELLAR MESSENGERS

I immediately perceived the true descent of this people, which does not appear of terrestrial origin, but descended from some of the inhabitants of the moon, because the principal language spoken there, and in the center of Africa, is very nearly the same. Their alphabet and method of writing are pretty much the same, and show the extreme antiquity of this people, and their exalted origin. I here give you a specimen of their writing . . .

Baron Munchausen

USING RADIO TO MAKE contact under Case II conditions, where the distance d to the nearest neighbor ranges from 30 to 300 light-years, becomes progressively harder as the distance increases. It is therefore worth looking for an *alternative method of making contact*. The problems to be overcome fall into three categories. *First, the geometrical*: statistics work against us as we search the sky with our radio telescopes, striving to have them aimed in the right direction at the right time and tuned to the right frequency. *Second, the fiscal*: cost in time and money to sustain a search program is substantial. *Third, the political*: there are seemingly insoluble political difficulties. How can we plan ventures whose success depends on availability of substantial funding decades into the future and which court failure if, at a particular future date when a crucial radio message reaches us from deep space, there is political turmoil such as war, revolution, or economic depression?

In the alternative plan to be unfolded here, we accept an unavoidable initial slowness in establishing contact, trusting that our patience in the beginning will be rewarded by a reliable, high-quality radio communication link later.

Exploration by Probe

There is a way to avoid or alleviate these three basic problems. The answer seems to be for the extraterrestrials to send a messenger. This need not be an individual of their race, although their biological engineering may well have proceeded to the point where a subrace of interstellar messengers has been bred. It would be completely unacceptable to us, but perhaps not to them, to breed brains in bottles, with eyes and arms but no bodies or legs; possibly one can do the equivalent with solid-state devices. One way or another, it is clearly possible in principle to pack an enormous amount of information into a modest *interstellar probe*, and to send it into the vicinity of a likely star. When the probe arrives, it fires a rocket and goes into circular orbit about the star in the middle of the habitable zone and in the star's equatorial plane, the plane where planets will be situated. The correct orbital size can be supplied to the probe before departure, but the orientation of the equatorial plane may have to be determined instrumentally, or at least checked, by the probe when it gets close enough to the star.

After settling into the approximate orbit where the inhabited planet should be (if there is such a planet around that star), the probe begins a program of investigation, supplementing its internal power source with stellar power in a way similar to, or better than, the scheme that has been so successful on Skylab and other solar-powered spacecraft. Since we know that the chances of finding technical life on the first star tried are very slender, as many tries should be made as can be afforded. Therefore, the probe will be the most modest design that can fulfill the functions we will soon describe. Even if no technical life is discovered, the probe will not have been expended without returns, for it can perform a variety of useful investigations. Among these will be the usual measurements carried out by our Mariner and Pioneer interplanetary probes on magnetic fields, high energy particle distribution, stellar wind, and color photography. In addition, it should be possible (by a telescope-imagetube-computer combination) to locate the planets and their satellites and determine their orbits. A considerable number of physical properties, including the size and mass of the planets, the composition of their atmospheres—and possibly biological attributes, such as the presence of life—should be accessible to automatic determination from the orbit of the probe.

Sophisticated Messengers

In my opinion the time will come when *we* will launch expeditions to the very nearest stars. It may not be in this millennium, but when solar system exploration is in an advanced phase we will outfit and launch our first unmanned stellar expeditionary probes. By that time, our sophistication will be considerable. I do not doubt that our first expeditions could be equipped with automated modules for landing on hard planetary bodies or for penetrating deeply into gaseous ones and studying either of them in detail. Such capabilities will result from our experience in exploring the solar system. All sorts of automatic tools for analyzing the soil and atmosphere, making biological cultures, and undertaking other complicated experiments will be available which, even in the first interstellar expeditions, will equal or exceed the ability of a live astronaut.

Systematic exploration by specialized probes concentrating on contact with advanced life will not be undertaken until exploratory expeditions to our immediate stellar environs (out to a few light-years) are well advanced.

The nearest superior community will be advanced in ways we cannot conceive of. Consequently, the sequence of events outlined above—exploration of one's neighboring planets, advancing to the whole planetary system, and then to the immediately neighboring stars—will already have been followed by some of the 10 million communities more advanced than we, which are distributed, albeit thinly, throughout the galaxy (under Case II, where our nearest superior community is 30 to 300 light-years away).

Therefore I believe that, under Case II, a program of dispatching specialized messenger probes will already be under way. Possibly such a program is currently being engaged in by our nearest superior community, whose home planet we will refer to as *Superon.* I will present the plan as though Superon is conducting the search and earth is an object of the search, since this is how I believe the situation is likely to be. But, if there has still been no contact with extraterrestrial civilizations by the time man has entered upon the exploration of his stellar environs, it will be time for *us* (rather than the extraterrestrials) to assume the more active role.

Budgetary Considerations

There is a natural tendency to suppose that because some civilization is incredibly advanced beyond ours—capable of performing technical wizardry unthinkable to us—it will therefore not be subject to budgetary discipline. A South Sea islander visited by European ships far exceeding his canoe in size and speed, replete with marvels such as steel axes, could be forgiven for supposing that his powerful visitors had unlimited resources. But in truth, we employ our resources to the limit no matter how vast they may be, and so it will be on Superon. Contact with other civilizations is not their only enterprise, and the director of the agency for sending missions to likely stars will have his slice of the budgetary pie allotted to him along with the responsibility of achieving the goal of detecting advanced life as soon as possible. He achieves this goal by launching the minimum equipment that will do the job, minimizing the cost per unit and thereby allowing the maximum number of likely stars to be probed. Thus we are led to ask what the minimum equipment is that would reveal the existence of technical life to a probe in orbit. The answer is a simple domestic radio receiver. The search procedure simply consists of turning the knob of a radio set to find whether there are any stations to pick up.

Detection

Although the probe is supposed to be in roughly the right orbit, it is not necessarily in the same part of the orbit as the prospective planet. Many months might elapse during which the star was unfavorably situated between the probe and the planet. But the simplest equipment will ultimately suffice to establish beyond question that radio and TV stations are present. This information would form part of the next report scheduled to be sent back to Superon.

It is quite possible that a probe arrived in our solar system when our radio transmissions had not yet begun. It would therefore be desirable for the probe to continue listening. We are sensitive to this point, because the earth could have been the target of a messenger probe as recently as a few decades ago, but even then we did not have commercial broadcast transmitters in operation. The probe may have dutifully reported this deficiency back to Superon and now be orbiting lifelessly somewhere in the space between Venus and Mars. Superon's best strategy, once an absence of technical life is reported, is to pass on to new likely stars. The number of new independent reports reaching Superon per annum is determined by the launch rate permitted by the budgetary allotment. If on first report a planet has no technical life, considerable time may elapse—perhaps several hundred million years—before technical life evolves. Rather than ruggedize its probes for such a vigil, Superon should allocate its money and effort in further minimum-design probes, allowing the spent ones to wear out in orbit.

Messenger Announces Its Presence

If all goes well, the time will come when a messenger probe has noted the presence of radio transmissions from a certain planet. Let's suppose that planet is the earth. The probe's next move is to let us know that it is there. Should it turn on a bright light, explode a bomb, or . . . why not radio to us? From our experience with spacecraft we know that a probe can get ample power from its solar cells for its transmitter and can maintain transmission for a considerable time. But what frequency will it use? In the case of a messenger probe, this is a non-problem, since the probe can rest assured that on any frequency where a transmitter can be detected there will also be, somewhere, a receiver! Therefore, the probe can choose any frequency which is already plainly in use. It is true that at least one receiver will be tuned in, and perhaps a large audience, but will they pay attention to an unwanted, interfering signal? Perhaps not, if the channel is used for taxis or walkie-talkies. But if the intruding signal came in on FM radio, a national TV network, a military frequency, or any one of many other channels, it would be identified as to direction and distance by many different countries within the day. After a further day of confusion as radio scientists compared notes by telephone, its orbit would be roughly known.

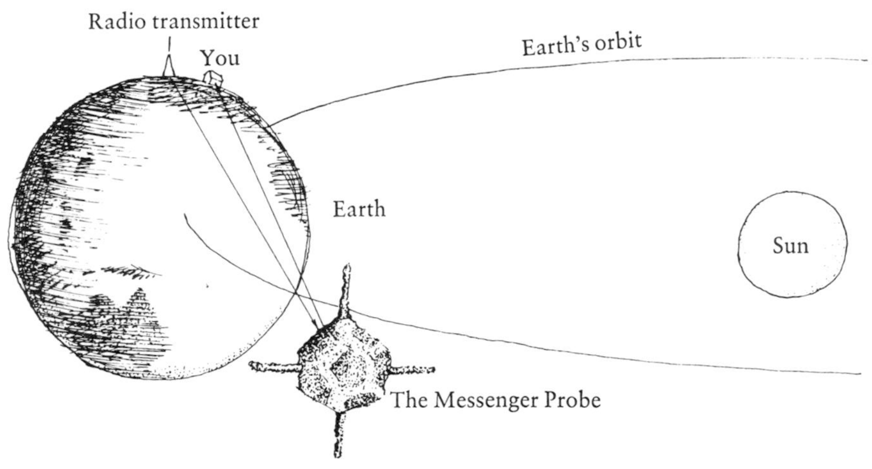

Fig. 10. If an alien probe near the earth picks up a terrestrial broadcast from a radio transmitter that you are listening to and then amplifies and retransmits it, you will hear first the direct transmission followed by a repeat after a time equal to that taken to travel out to the probe and back. The general effect will be like an echo.

Up to this point, no message need be coded into the probe's radio transmission—mere transmission would itself suffice. A simple procedure would be for the probe to amplify and retransmit the same TV program or military communication it was receiving. Its signals would then have the appearance, to us, of echoes exhibiting delays of seconds to minutes depending on its distance from earth. For instance, if we were listening to the radio, each word would be heard twice, first by direct transmission from the station and then again a little later via the probe. Let us assume that the probe does adopt this initial procedure.

Acknowledgment: Letting Them Know We Know

At this stage the probe knows we are here, and we know it is there. It is now our turn to let it know *that we know* it is there! Special transmissions must be arranged with the station which the probe is echoing. Changing the transmission to short phrases separated by quiet intervals in order to remove any overlap of each phrase and its echo, we would begin to do to the probe what it is doing to us, namely, to repeat the echo. The probe will detect that the character of the program has radically altered to one that is periodic and repeating with a period that could only arise if the probe was somehow influencing the transmitter. Who knows how fast it would notice the alteration? I would say promptly. In case any doubt lingered in its mind, to use an anthropomorphic term, it has a general purpose maneuver preplanned for this verification: by merely ceasing to echo, and noting that we do likewise, it *knows* we are onto it!

Messenger Sizes Us Up

At this point, a number of possibilities present themselves. Our project director would be under some pressure to induce the probe to shut down or move to another frequency, so that the regular business of the channel could resume. But of course the probe has the initiative, and, as it knows very little about us or our equipment and must not lose this first precarious contact, I think it would begin to interrogate us on practical housekeeping matters such as: 1) how sensitive our equipment is, 2) what the bandwidth of our equipment is or how fast we can receive, and 3) whether another frequency would be better in view of technical or political matters that we know about but it doesn't. It also needs to guard against the loss of its message when, because of the earth's rotation, it sets below the horizon of whomever it first contacts. For all it knows it may have to be prepared for sporadic meteorological phenomena such as lightning, sandstorms, or sunspots—all of which can interrupt radio communication.

To test the sensitivity of our equipment it merely weakens its echo; we repeat back, it weakens it further, we now realize what it is up to; and when the signal level is weakened to a point that is uncomfortable for reception, we cease to repeat. Having obtained this information, the probe shifts frequency slightly; we follow. It shifts a little further. If we are using a commercial broadcast transmitter we will have great difficulty in following because such transmitters are not designed to go beyond narrow limits that are strictly prescribed by the Federal Communications Commission. The probe can thus, with our assistance, determine technical parameters of our radio transmitter rather simply. As soon as we can bring a variable-frequency transmitter into action, we ourselves can take the lead in changing the frequency slightly and leading the probe off to a frequency, within the probe's range, that is more suitable for us.

Political Obstacles to the Probe's Free Communication

It would be naive to think that an alien probe could count on delivering its message freely to the inhabitants of the earth. Harvard biologist and Nobel Laureate George Wald believes "disaster lurks in the stars": that contact with another civilization would produce "the most highly classified and exploited military information in the history of the earth." Although I agree that any agency able to do so would keep the message secret regardless of its content if the opportunity presented itself, I doubt that there would be such an opportunity. All radio transmitting on earth is subject to strict political regulation with heavy international and military overtones; it is common for terrestrial governments to use jamming to prevent disturbing broadcasts of foreign origin from reaching their nationals. The probe's message will be disturbing. Therefore the probe must be prepared with sociological resourcefulness to avoid being trapped into secrecy, to avoid exclusive relations with one power that would invite its being shot down by a rival power, and to avoid being jammed by a minor power.

Political matters of this type are not generally considered in discussions of making contact by radio, but they could be very real problems determining the success or failure of the attempt at contact. Thus it is of direct importance to visualize the course of events following the probe's announcement of its presence. We will begin with the point where control passes into the hands of a technically equipped radio scientist or engineer. Let's call this scientist the project director. He could belong to any one of a number of adequately equipped national laboratories or universities in several countries. Wherever he is located, the rotation of the earth on its axis is going to cause the probe to set below his horizon within hours. He will have to arrange to hand over control to another longitude on earth where the probe will be above the horizon. NASA offers suitable facilities—equally spaced around the world in the United States, Australia, and Spain—with excellent intercommunication links. Given ample warning, NASA could no doubt organize 24-hour surveillance of the probe with accredited international representation. However, if the probe was first discovered in France or England, and in the first day of confusion several nations participated in interrogating the probe, organization would be in a shambles—though I doubt that the probe would be confused. After all, such chaos is one of the obvious complications that it would have expected. I think the probe would have a plan aimed at identifying a competent worldwide entity to receive its message. For example, the probe could enlist NASA's cooperation promptly, without the delay that would ensue if the earth people had to reach an international agreement (on who could transmit and who could not), simply by reducing signal level until respondents not having large antennas, sensitive receivers, and worldwide interconnection dropped out. The organization that remains in this game is the one the probe wants to deal with. The nations would then quickly accept an offer to form international teams to man the NASA sites, I would suppose. If, however, some nation insisted on talking to the probe independently, it would be necessary for NASA to defer to that nation for part of the day or to invite jamming by that nation by failing to defer. The contingency plans of the probe would need to provide for rivalry. Perhaps with further thought we could arrive at a wise course of action for a probe beset by two suitors, one trying to stifle the other.

Such political problems may be more serious than the scientific ones. As we have seen, much give-and-take on quite a high technical level could take place between ourselves and a preprogrammed messenger probe—very swiftly, even before a language of communication was set up. But the messenger probe does have a message; now we can speculate as to what it might be.

The Message

In my opinion, the message will be in television. Television is like sign language; although you and I may not speak the same language, we can exchange ideas through signs or pictures. Geometrical shapes furnish a means whereby we learn each other's language. The words in the dictionary that can be defined by drawings probably run into the thousands and those that can be defined by animated drawings are many more. Not only nouns, but many adjectives, adverbs, and verbs

can be depicted through television. Other words are harder, but if one had a dictionary and already knew a few thousand basic words, one could interpret some of the more difficult ones.

Until we set up a common language, television also permits us to ascertain quickly the answers to questions of basic importance, such as where the probe came from. The probe, too, would like us to know this without delay. To speculate a bit, I will assume that television is what we are going to see, and that the first picture will be of a constellation of stars, familiar to us, followed by a zooming in on the home star. At one time I thought the picture of the constellation would have one star winking on and off like an electric sign, but anyone who can make an animated movie to do that can just as readily simulate a zoom lens. The zoom lens technique is a very quick way for a foreign probe to tell us which star it came from without knowing our name for the star, or our coordinate systems, or anything about our language. You might also want to know how we are going to get our TV screen synchronized to its system of transmission. If the probe keeps running its program until we get the number of lines per frame and so on worked out and arrange a set to display it, then we can report our readiness by repeating the program back. Alternatively, since it has been listening in on our TV transmissions, it might oblige us by adopting one of the numerous standards in use on earth.

Now that we know the home star, the probe will zoom in further until the star grows into a visible disc, perhaps with starspots. From their motion we will know the axis of rotation. The planetary system will then be displayed, and at last we will zoom in on Superon, the home planet. Clearly a fantastic travelogue lies ahead, and it is well within the capacity of a modest probe to execute the simple steps required to display this information. After this brief preview, there is some rather serious business to attend to, namely: the messenger probe must convey to us the schedule of listening that is being observed on Superon and the frequency being used—this is rather urgent. Afterward the probe is dispensable, but until the schedule and frequency are conveyed the probe's mission is incomplete. I believe this job can be accomplished with pictures, but I surmise the programmers on Superon will surprise us by the brevity and clarity of their particular method of conveying this vital information.

Language Barrier?

The nature of *our* direct transmission to Superon is a matter of taste. Obviously we can send zoom movies to match those the probe will have shown us, but much more interchange can take place with the probe while our direct transmission is being readied. I believe the probe can learn our language in printed form quite readily, working with an animated pictorial dictionary that we furnish for its computer memory. At first its expressions might be quaint, but there is no reason why its compositions should not be televised back to it in corrected form. To be sure of getting a point across, the probe can do what we do—say it again in different words—and if we don't understand, we can question.

Knowledge of our language will enable the probe to tell us many fascinating things: the physics and chemistry of the next 100 years, wonders of astrophysics yet unknown to man, beautiful mathematics. After a while it may supply us with astounding breakthroughs in biology and medicine. But first we will have to tell it a lot about our biological makeup. Perhaps it will write poetry or discuss philosophy. Perhaps the messenger knows how the universe started, whether it will end, and what will happen then. Maybe the probe knows what it all means, but I wonder. . . . I think that is why Superon wants to consult us!

The Limit

In time, the probe's store of knowledge will be used up, as it is only a modest probe. Presumably the computing part need only be the size of a human head, which is, we know, large enough to store an immense amount of information. Meanwhile our transmission to Superon will have commenced. One might ask whether it would be better to use our language or Superese, which we could learn from the probe. Now while I think the probe could learn a functional form of our language, I don't think it would be practical to teach it to the people of Superon. The continual checks and confirmations, corrections, and repetitions that are possible between ourselves and the probe resemble face-to-face encounters between humans meeting on language frontiers and would be ruled out by the round-trip delay to Superon and back. It is true that we could transmit our pictorial dictionary followed by text, and hope for the best; if the probe met an untimely demise that would be the only course. However, by the time the probe is well advanced in its mastery of our language, it will possess an on-board translation program that converts our language to Superese. Therefore we can copy its program and either transmit it to Superon or translate into Superese locally before transmission. Perhaps the probe would advise us as to our degree of success.

Mission Accomplished

From what has preceded we can see that the messenger probe scheme for contact overcomes dependence on terrestrial socioeconomic and political stability over centuries, circumvents the problem of determining the proper frequency, and is well adapted to its primary mission of detecting our existence and announcing the location of Superon. If the probe accomplishes no more than this, it has achieved the initial detection that seems so fraught with difficulty if attempted by direct radio; in fact it does better, by reporting to a home base which is ready and waiting to receive the report!

The Galactic Club

Meanwhile, back on Superon, much time has elapsed since the messenger probe set out, so receipt of the probe's report is dependent on the very kind of political and economic stability that Superon should avoid relying on. But this is a relatively mild dependence for several reasons. First, the schedule and frequency of transmission as well as the report's direction of arrival are known, in fact, have been known well in advance. Second, the receiving equipment exists. Third, no financial outlay of great magnitude is required. And, fourth, the report will be interesting because of the new data it contains about another planetary system—even if it does not carry the spectacular news that Superon is ultimately seeking.

As a protection against assorted mishaps, such as failure of the report to reach Superon, the probe can announce to us the location of other,

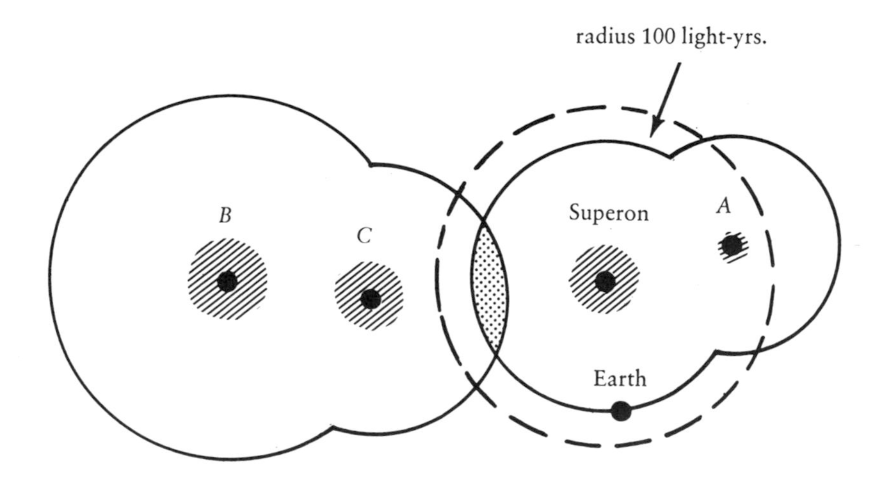

Fig. 11. Spheres of exploration expanding around intelligent communities in the galaxy. Superon, our nearest superior neighbor, has just contacted the earth. Some time ago, it found technical life at *A*, which is now making its own sphere of exploration.

more remote members of the Superon chain of communication. For there will likely be a galactic club, whose members are experienced at finding developing communities such as ours and inducting them into the galactic community. Each will have responsibility in its own sphere of influence and will engage in an ongoing program of launching messenger probes, at an annual rate appropriate to local priorities, in an endeavor to comb their unexplored frontiers for new technical life. An idea of scale may be obtained by referring to *Fig. 11*, where four superior communities—*A*, *B*, *C*, and Superon—are shown. The shaded circles, ranging in radius from about 10 to 25 light-years, represent the volumes of space around those communities that have been rather closely inspected by well-equipped expeditions; the larger circles show the limit now reached by the messenger probe program. Superon has just reached the 100 light-year range, finding the earth. Some time ago it located technical life at *A*, which was inducted into the chain, is now a superior community, and is participating in the messenger probe program as indicated by the circle showing the progress of *A*'s own exploration. Two other members of the chain are *B*, which is relatively old and has explored more space, and *C*, which was inducted by *B* long ago. Superon is in contact with *C*, not as a result of probe exploration but because Superon became independently aware of the existence of *C* when news of its discovery was relayed via links in the chain not shown in the diagram because they are not in the plane of the paper. Had this not been so, an interesting situation could have arisen in the lens-shaped region of overlap between Superon and *C*, where each could have had messenger probes in the field. (It is a whimsical thought to contemplate two automatons, both far from home in the reaches of space, exchanging notes about their builders, though I admit it is not a very likely encounter.)

Many localities remain unexplored in the vast crevices between the expanding spheres. For example, by the time Superon expands its probed sphere from the 100-light-year radius shown to the 125-light-year radius indicated by a dashed line, it will have doubled the number of stars visited. Therefore, even this apparently modest expansion will require much time. Of course, exploration is helped as new communities such as *A* take over part of the work.

The Time Scale of Interstellar Dialogue

If the messenger probe plan is so good, why do the scientific publications relating to extraterrestrial contact refer mainly to radio? I think the answer is that if a superior community averages one launching per year, 1,000 years will pass before enough probes have been launched to

The Solomon R. Guggenheim Museum, New York

cover the likely stars within 100 light-years. When the first technical life is discovered, decades will elapse before the probe's report filters back. In addition, we have to consider the travel time of the probe to the star's environs. Depending upon the size of engine the probe uses, the travel time could be kept down to centuries or even decades. But a community that is prepared to wait it out for 1,000 years does not need to hurry. Perhaps 10,000 years travel time would be reasonable; it depends on a trade-off involving reliability in transit, cost per unit, number of launches per annum that can be afforded within the budget allocation, and the maximization of the probability of success. This is indeed a long-term project! (By the way, note that interruption of the launch program does not affect the chances of success of probes already launched; the plan is tolerant of diversion of resources to urgent priorities.) Of course, the human life span being what it is, we are reluctant to contemplate programs that stretch over centuries; we have to realize that interstellar contact is not contact between individuals, but a contact between civilizations. This is a slightly depressing thought for action-oriented people. The arrival of a probe would be exciting. Nonetheless, the individual who directs the launching, be it a probe or a radio signal, has to face the reality that it will not be he who receives the answer.

Are Messenger Probes Durable?

Technical difficulties have been raised about the design of messenger probes. The first is whether a probe could survive the battering by radiation and particles in space for a long duration. I.R. Cameron presented information bearing on this question in *Scientific American*, July 1973, where he described methods for estimating the erosion rate of meteorites. One might think it impossible to tell whether a rusty meteorite found in the ground had lost any of its surface material by erosion in space. But this can be done, Cameron explains, through methods of determining the age of the meteorite beginning from the time when it was first exposed to radiation in space (presumably at the time of collision of two parent bodies).

As this advanced technique comes to be applied to the abundant meteorite material that is available, much will be learned about conditions in space over the eons. Meanwhile, indications are that the amount of erosion over a million years ranges from 0.2 millimeters to about a centimeter, apparently depending on the particular composition of the meteorite—and no doubt on where in interplanetary space it spent its exposed life. These results show that quite a thin coating on a messenger probe will amply protect it from the cosmic radiation for the duration of its mission.

A second difficulty with probes is how to make reliable electronic equipment capable of functioning after many years of lying idle in space. A comparable terrestrial requirement is found in transatlantic submarine telephone cable that contains many amplifiers built into the cable, each containing many vacuum tubes (not transistors, interestingly) which are required to work as a viable whole for a full 20 years of submersion (and hopefully longer). Even though not a single one of these components had been subjected to a 20-year test, engineers were able to simulate such a test with confidence. Therefore, although more recent devices such as transistors have not been tested over long time intervals, it does not mean that their behavior cannot be foreseen. The deterioration of a transistor depends in a calculable way on its temperature; thus it presumably can be determined whether the cold storage available in space (close to absolute zero) is sufficient to preserve transistors for the required length of the mission. I think there will be a way of designing around this problem, even if it entails producing some of the devices within the probe as it approaches its destination.

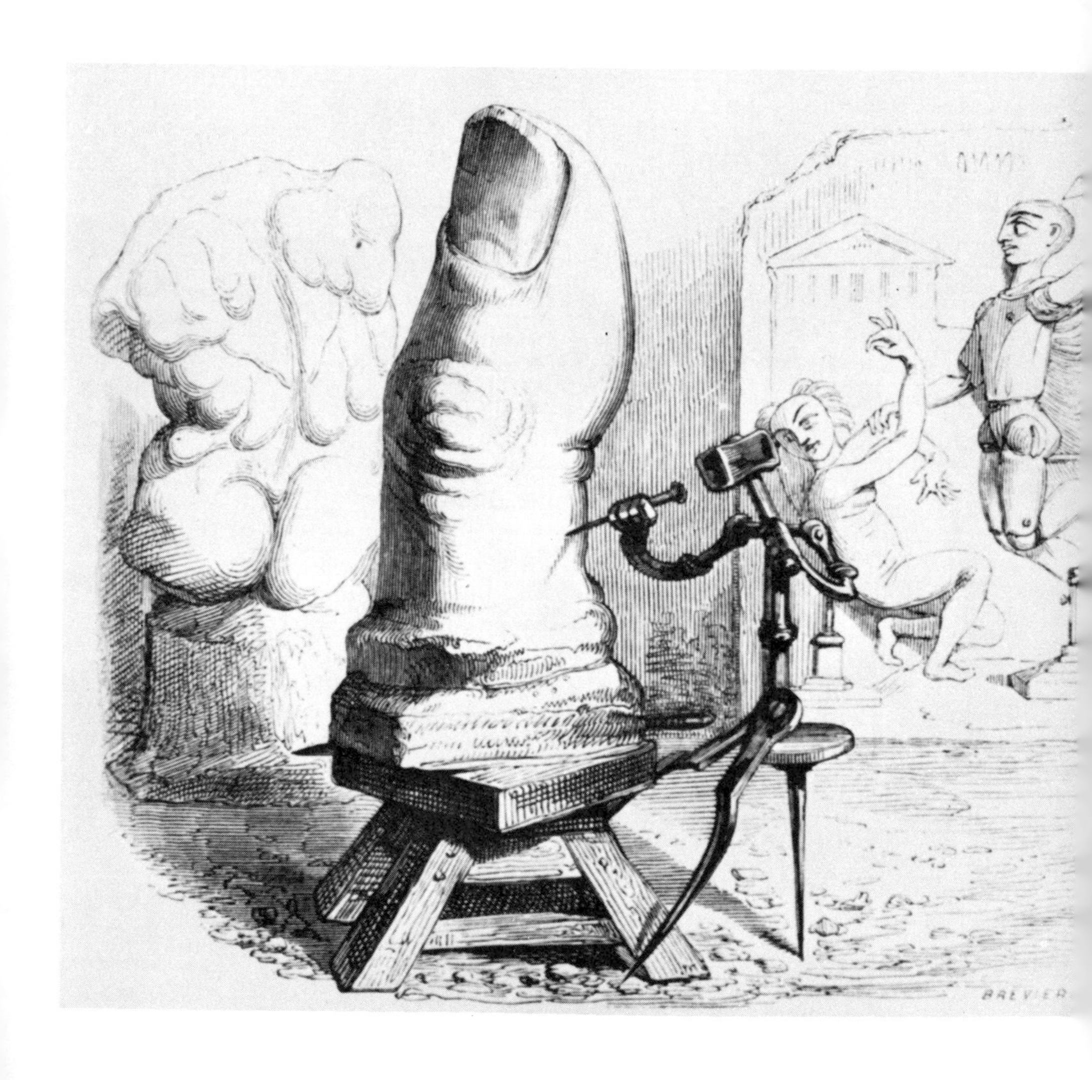

CHAPTER NINE

WHAT DO YOU SAY TO A NONHUMAN?

Never look askance at a tree or a cabbage. Its friendship you may sometime need.

David Starr Jordan

MANY PEOPLE FEEL that we would have little to say to an extraterrestrial being; the very thought of an inhuman intellect clothed in a strange body is enough to give some people the shivers. How could we ever understand the mentality of such a creature? I don't think we ever could. But, where we have overlapping spheres of interest we will be able to deal with an extraterrestrial being.

A nonhuman alien who is able to navigate through celestial space will be able to exchange lists of astrophysical data with us to mutual benefit and will share an interest in such chemistry and physics as we possess that may be new to him. Since many of our current problems —peace, energy, pollution—depend upon as yet undiscovered physics and chemistry for part of their solution, our interest is more than merely academic. The whole subject matter of pure mathematics will surely fall within our overlapping spheres of interest. But we will not know and may be unable to conceive of the values that he attaches to these items. Conversely, we may be unable to convey to him what human emotions such as anger or love mean to us. Perhaps he will be able to follow our explanation in an intellectual way but be unable to feel or identify with experiences that for us are deep-seated.

A good parallel to contemplate in preparation for any impending meeting of races is the relationship between a man and his dog. Close as we are, both being mammals, we are worlds apart. Yet we have many things in common: chasing one another, feeding or being fed, petting or being petted, showing affection. But to discuss science or poetry or your love life with your dog would be pointless; these things we do not have in common. Some such division of interest will apply to each extraterrestrial community that we are destined to encounter.

Forms of Intelligent Life

The marvelous range of life forms on earth, each molded by its place in life, warns us to prepare for at least as much variety among the intelligent entities occupying the diversity of niches offered by the whole universe. To the extent that our contact with alien life is limited to exchanging messages, the precise physical form of our correspondents seems unimportant. But their form is not without interest and could have a bearing on their outlook, thus influencing the content of messages and the degree of overlap of our spheres of interest.

It is by no means likely that other intelligent life would resemble man physically. We have only to consider a planet where gravity is several times stronger than on earth to see that evolution would favor creatures that crawled flat on their bellies like crocodiles. Instead of an arboreal ancestry, such creatures might have found their antecedents in swamps, where gravity could be counteracted by buoyancy and then, instead of taking to the plains and becoming upright walkers, as our ancestors did, they might have moved out from their place of genesis along waterways, natural and artificial. There are many other ways in which a planet could differ from earth and force development of radically different life forms.

It is conceivable that intelligent life need not be based on a society of individuals that reproduce sexually, spend much of their life span maturing, and are replaced by new models every 70 years or so. We also cannot assume that our size is representative.

Perhaps there could be intelligent scum. After all, the size of the human individual has practically nothing to do with the size of his pyramids, cathedrals, oil-tankers, rockets, and other products of technology. Had we been 12-foot giants or 3-foot dwarfs would it have had any influence on the scale of such undertakings? I do not see why scum composed of single-celled plants, living on a water-air interface, possessing no resistance to wind or weather, could not attain technological control of the environment.

One cannot propose the train of development that would lead to intelligence in colonies of scum but, assuming the end result to have been reached, one can at least note some of the conditions that would need to be met. The extreme vulnerability of an ephemeral scum means that the intelligent scum must encapsulate itself in a way affording it mechanical and thermal protection from the environment. We see that the human brain is a convoluted assembly of cells distributed in layers, something like a silk sheet stuffed into a bone bag. I would expect the scum to find a way of rolling itself and its substrate into a ball and acquiring a protective coat. Perhaps this ball would be about as big as a human head, but the contents, if evolved from a monolayer of unicellular organisms, would not be an individual as we understand the word, but rather a whole tribe!

Because of the modest abilities of unicellular creatures, cooperation and specialization would be required on an intimate scale, which proximity would facilitate. In the course of time, the cells on duty at the mouth of the bag might be the only ones to retain sensitivity to light. These sentry cells would be evolving toward eyes. If circumstances so dictated, it does not seem at all implausible that originally unicellular individuals could specialize so as to perform the functions of different tissues and organs. "Slime-molds do it all the time," Dr. Lewis Thomas tells us in *Lives of a Cell*, referring to free-swimming myxomycetes, which come together in numbers and fuse into a single slug.

The converse of intelligent scum is one enormous entity that functions as the equivalent of a whole civilization. In Fred Hoyle's fantastic story *The Black Cloud*, an intelligent cloud resides in interstellar space. Its *brain* consists of layers of suitable molecules that it synthesizes itself and assembles on a solid substrate. To recharge and cleanse this material a flow of gas containing the right substances performs the functions of blood, and the waste products are removed by a filter equivalent to a kidney. A pump acting as a heart causes the blood to flow. Many units of this kind are distributed throughout the cloud—interconnected by radio instead of nerves. As they are manufactured by the cloud itself, spares can be kept in case of failure and the total brain capacity can be enlarged and specialized as desired. The lifetime of the cloud is essentially unlimited, but it occasionally runs out of energy and approaches a star whose energy it uses for building up needed food molecules.

Even *intelligent machines* could be members of the galactic club. The builders of the machines might have been faced with impending extinction by a steadily intensifying ice age (which, as biological organisms, they did not expect to survive) and could have made the machines out of a desire to preserve their culture. Alternatively, some community may have launched intelligent machines to investigate or colonize distant parts of space. No doubt the designers of a successful intelligent machine, one intellectually indistinguishable from its creators, would entrust a few fertilized eggs on ice to the care of the machines, so that,

should conditions permit, individuals might be recreated in their original form.

Were we ever to contact intelligent machines by radio, it would be difficult for us to know that they were merely machines. But would there be a distinction? Like us, they would be a product of the physical world and therefore part of Nature. What we do not know for sure, since it has never yet been demonstrated, is whether an artificial machine can *think*, and until this has actually been done, many will harbor a lingering doubt.

Alien Life Styles

We are familiar with the different life styles of human societies and with the difficulties that can arise when one group tries to communicate with another. But at least we are all human. Different human societies share approximately the same life span, imposed by a combination of accident, disease, and old age. If, on another planet, infirmities of disease and old age were much less prevalent, accident being the chief cause of death and the life span averaging 200 years, life styles would be radically altered. There would be a small but significant proportion of 1,000-year-olds, including a sprinkling of patriarchs playing dominant roles such as are not seen on earth. Under conditions of zero population growth, child-bearing years would become a brief interlude, and if the intelligent creatures were descended from wild ancestors that bore litters rather than primarily single offspring, then the average family would never be called upon to have children at all. A human spokesman entering into discourse with a member of such a society could be excused for feeling like the proverbial fish out of water.

We can visualize the consequences of encountering an elderly society. However, an alien society might have aspirations so different from ours as to be very difficult to comprehend. Reasons can be given for thinking that aspirations can change profoundly. Hitherto, human history has been one of expansion, built on an incentive of gaining an increased return from exertion. Where effort does not lead to reward we have decay. But parts of the world are reaching, or have reached, the limits of expansion and face stagnation, catastrophe, or conscious stabilization. But we do not yet see what goals one aspires to in a stabilized (no-growth) society. Preoccupation with technology might cease after some stage in a community's development and intelligent life might become contemplative or turn more to philosophy and the arts. If civilizations like this exist, they could be hard to detect. The ones that are out striving to conquer space are those that we will be dealing with. Perhaps conquest of space will spur man forward after the earth is filled.

Miró

CHAPTER TEN

HAVE THEY BEEN?

Ignorance is the most delightful science in the world because it is acquired without labor or pains and keeps the mind from melancholy.

Giordano Bruno

MAN AWAITS THE FIRST message from space with characteristic impatience, which could account for announcements made by various individuals from time to time that they have received unearthly messages —telepathically, in writing, or even directly—from extraterrestrial visitors. The inconsistencies among the numerous reports of this kind make it likely that many of them are wrong. It is too much to suppose that messages from Venus, Mars, Jupiter, and Andromeda, to name only a few, have all arrived in recent years; it may be presumed that a certain fraction of these reports, perhaps most, are psychologically based.

It is worth addressing ourselves to the possibility that the first message arrived some time ago. Suppose an interstellar messenger probe, on arriving at its destination, found no signs of intelligent activity but did discern signs of life that might in due course evolve to intelligence. It would make sense for the probe to leave a visiting card of sorts. What would make a conspicuous and durable sign?

A Cosmic Visiting Card

One possibility would be to leave the messenger probe in permanent orbit in the solar system, trusting that it would ultimately be detected, perhaps by radar or by the great radio telescopes which ceaselessly record the heavens. A very good chance exists that it would be found by a civilization, such as we now have on earth, within the first few decades of the exploration of interplanetary space by astronauts. Such a passive probe may well be in orbit now, having long since expended its ability to communicate by radio, battered by long bombardment of interplanetary dust, solar wind particles, and other hazards of space. Locating a wreck in this condition could be difficult but, when located, it could be a treasure chest of information. Studying the equipment of an alien race has fascinating potential. Lessons of metallurgy or materials science could exercise scientists for years. Think of today's sophisticated materials, such as the artificial ruby (on which the generation of red laser light depends) and the silicon and germanium crystals (which have permitted the advance of solid-state electronics); they would have baffled a 19th-century metallurgist, chemist, or physicist. He would not have had the tools to recognize their essential features. But if he knew that the material was taken from equipment which had come from another star, he would have at least locked part of it away for future study, just as we saved moon rocks. Thus, as the hulk of a worn-out probe would itself constitute a good visiting card, why not just allow it to wear out in space? Alternatively, why not crash the probe on the earth after failure to establish radio contact? The debris could be found sooner and more easily than an orbiting object—providing the debris was distinctive. The probe would have to guard against burning up in the earth's atmosphere, crashing into the ocean and sinking, or burying itself in the soil. For durability, it would have to guard against being eroded away by the elements, silted over, or concealed by vegetation. When we think of the Mayan monuments of Central America that have been devoured by the jungle and the vanished civilizations of Africa and Asia, we realize that even cities can be obliterated by the elements in a millennium or two.

Consider this. Suppose a container was filled with small, stainless steel spheres and allowed to fall into the earth's atmosphere. When the container burned up, or exploded by design, the spherules would fall in a swathe across the land, leaving a trail the way hailstones do. Some might fall into the sea, others into jungle, but some might fall on land favorable to their preservation. All it takes is a modest amount of distinctive material spread over a wide arc. Such things do exist on earth: tektites.

Tektites are glassy buttons found in many parts of the world, any particular variety (of which there are many—differing in color, composition, and shape) existing over a more or less definite geographical area. For example, one variety known as australites, which are a few thousand years old, can be picked up on an extended zone of stony desert running across Central Australia. Of course, small objects strewn over an arid desert are likely to remain in evidence for a long time; it is impressive to note that Texan tektites, tens of millions of years old, are still in evidence after the profound geological and climatic changes that have occurred there over perhaps 50 million years. The tektite story demonstrates that a swathe of small artifacts could leave lasting traces.

However, it has been determined (through artificial production of tektites by firing glass pellets in wind tunnels into an atmosphere simulating that of the earth) that the tektites originated in the impacts that formed lunar craters. Occasionally, the splash from a lunar impact ejected a jet of molten glass that escaped from the moon and fell on the earth along a strip of ground.

Other ideas have occasionally been proposed as to how attention might deliberately be drawn to a past visit: building pyramids, raising obelisks, planting forests, and so on. It would not seem too difficult for an extraterrestrial visitor to leave an unmistakable sign that would be durable for many thousands of years.

But the first message might have arrived 100 million years ago or more. In such a time period ice ages come and go, mountains wear away, seas become dry land, and dry land submerges. Deserts form, forests advance and recede, and evolution brings new species and eliminates old. Physical signs of such antiquity might be sought in the fossiliferous strata of the earth's crust and yet never be found. What about a biological sign? If alien seeds or spores were dropped on the earth, their progeny might be with us today and announce to us, by their distinctive makeup, that they do not share a common origin with our other life. If such an event did occur, it could not have been geologically recent. The common chemistry of *all* forms of terrestrial life indicates a long period of coexistence on earth.

Seeding Life

Could the earth have been *seeded* originally with the living matter from which all existing terrestrial life is thought to descend? At present it is widely contemplated that life on earth originated spontaneously (from molecules that formed naturally in the atmosphere and seas, developing through slow chemical evolution). But, as a gesture by a visitor

7-33

from outer space who found a lifeless earth, what could be finer than the initial seeding that would lead to the richness of living forms we have on earth today? A less romantic version of this theme, presented by the well-known Cornell astrophysicist Tommy Gold, is the Gold Garbage Theory, according to which life on earth could have originated in the waste products dumped long ago by space travelers visiting the earth.

I do not believe there is any way of demonstrating that life on earth did not originate by deliberate or accidental seeding by space visitors. There is the additional possibility of seeding from afar without an actual visit by space people, as well as the possibility of cosmic panspermia, which means that the seeds of life are distributed throughout space.

Recent Visitors?

As for suggestions that the earth has recently been visited by beings from space, no scientific evidence exists that commands any wide assent. There are indeed current controversial proposals seeking to interpret various myths and legends as accounts of visits from outer space. Stories about supernatural beings are so universal in human societies that there is no lack of rich materials on which to base such musings. Let us explore some of these suggestions.

CHAPTER ELEVEN

THE CHARIOTS OF VON DÄNIKEN

Let us just say that there are two sorts of poetical minds—one kind apt at inventing fables, and the other disposed to believe them.

Galileo, *Dialogue Concerning the Two Chief World Systems*

> The gods of the dim past have left countless traces which we can read and decipher today for the first time because the problem of space travel, so topical today, was not a problem, but a reality, to the men of thousands of years ago. For I claim that our forefathers received visits from the universe in the remote past. Even though I do not yet know who these extraterrestrial intelligences were or from which planet they came, I nevertheless proclaim that these "strangers" annihilated part of mankind existing at the time and produced a new, perhaps the first, *Homo sapiens*.

THIS PASSAGE from the introduction to Erich von Däniken's *Chariots of the Gods?* conveys the views that have given rise to much animated discussion among von Däniken's 12 million readers.

Among the physical traces left by the "gods of the dim past" are an inscription on the wall of a Mayan building at Palenque, Mexico (reminiscent of an astronaut in a rocket from which a jet is escaping), Easter Island statues, an iron pillar hundreds of years old yet rust-free, a Babylonian cuneiform tablet of eclipses, and the Great Pyramid of Cheops.

Pyramids Built by Gods?

The Great Pyramid of Cheops in Egypt is a massive structure 480 feet high (now somewhat reduced by erosion), with a square base approximately 750 feet long, and contains over 2 million blocks of limestone averaging about 2½ tons each. While not as tall as modern skyscrapers it contains vastly more material. A comparable modern project would be the Hoover Dam, which contains about 6 million tons of concrete. As the pyramid was constructed in the 4th dynasty, around 2600 B.C., people have justifiably been impressed by this early undertaking. Egyptian murals of various ages depict masons at work, showing their tools and methods of dressing stone, as well as teams of men pulling on ropes as they haul loads (weighing many tons) mounted on wooden sledges.

Wheeled vehicles, sledges on rollers, and gigantic levers are portrayed. Earth ramps (inclined planes up which stones were hauled) were used as the equivalent of modern scaffolding for constructing stone walls. In the construction of a pyramid, the body of the pyramid itself could be utilized to support a ramp that spiraled around and up the sides. The workers were slaves, shown in the murals to be under the supervision of guards with swords and batons.

Von Däniken believes it is fantastic to suppose that the Egyptians built this pyramid themselves, but is vague as to whether "gods" were responsible for *all* the pyramids or simply lent a hand on the largest one. Explaining why the Egyptians were not responsible for constructing Cheops, von Däniken writes:

> I shall be told that the stone blocks used for building the temple were moved on rollers. In other words, wooden rollers! But the Egyptians would scarcely have felled and turned into rollers the few trees, mainly palms, that then (as now) grew in

Egypt, because the dates from the palms were urgently needed for food and the trunks and fronds were the only things giving shade to the dried up ground. (Bantam, p. 75.)

Is this true? The records from the 4th dynasty tell of mining expeditions to Sinai where copper could be found in the south; timber trade with Byblos (on the coast of Lebanon near Beirut), a consignment of 40 vessels with timber being recorded; and stone quarrying up the Nile at Aswan for granite and in the Nubian desert (even farther south) for diorite. This demonstrated ability to travel to sources hundreds of miles away and transport massive cargoes explains why it would *not* have been necessary for the Egyptians to sacrifice local palm groves to make rollers; importing cedar logs was well within their capacity.

Addressing the supposed construction of the pyramid by Egyptians, von Däniken expounds: "Several hundred thousand workers pushed and pulled blocks weighing 12 tons up a ramp with (non-existent) ropes on (non-existent) rollers."

Yet, rope was of fundamental importance to Egyptian construction work, permitting, as it did, a large team of men to fan out over the necessary space in front of the sledge they were drawing. (See relief.) Ropes made of reed, date-palm fiber, and other materials have been found dating to prehistoric Egypt, around 4000 B.C. In the caves at Tura, where the fine limestone blocks used for facing the pyramid were quarried, ropes 2½ inches in diameter have been found—composed of three strands, each strand consisting of about 40 yarns, each yarn spun from about seven papyrus fibers. Quantities of rope were also used for rigging the vessels plying to Lebanon and up the Nile. To one familiar with Egyptian cedar and rope, quantities of which may be seen in museums, von Däniken's unsupported assertions are not persuasive. In fact, his failure to mention the cedar trade or the museum samples of heavy rope and cordage indicates that he is not giving up-to-date knowledge about the subject under discussion.

Hidden Treasures?

Since the appearance of *Chariots of the Gods?* von Däniken has written *Return to the Stars* and *Gold from the Gods*, which together have sold over 12 million copies in 32 languages. This last book upholds his view that spacemen visited the earth long ago. Accompanied by archeologist Juan Moricz and Franz Seiner, he allegedly descended into an astonishing world of subterranean galleries hidden beneath the tropical jungle of the Ecuadorian province of Moreno-Santiago. Metal tablets found in these caves support von Däniken's views of previous visits from space. The book gives details of the underground furnishings, wall carvings, metal plaque library, birds, snails and crabs, conjuring up a remarkable scene. Yet, when the German magazine *Der Spiegel* sent an expedition to follow von Däniken's footsteps through South America and to interview the personages in the book, the findings repudiated many of his assertions. According to von Däniken's Ecuadorian guide, von Däniken did not actually descend into the galleries at all, but rather pumped the guide for all the information he had. Von Däniken later admitted that his itinerary in Ecuador did not allow him the time necessary even to reach the location of his purported visit so vividly described.

Space People?

More recently von Däniken has investigated statuettes made by the Ainu of northern Japan. From their rather large heads and cumbersome form he recognized them as visitors to earth clad in helmets and spacesuits. After articles by A. Kazantsev as well as von Däniken appeared in the Soviet publication *Za Rubezhom* (*Abroad*, No. 9, 1970), the Soviet ethnographer S. Arutyunov refuted the assertions of von Däniken in an article entitled "These are not Cosmonauts" in *Znanie—Sila* (*Knowledge is Strength*, No. 9, 1970). Arutyunov explains:

> In the curious decoration adorning the Dogu statuettes, a vivid imagination can quite easily discern spacesuits with sealed helmets, goggles with slits for the eyes, and filters for breathing . . . little rivets, buckles and hatches for inspection of the mechanisms on the shoulders and on the back of the helmet. If the Dogu statuettes were to appear all of a sudden among the other antiquities of Japan, in spacesuits, there would be a basis for such an assumption. But the point is that the statuettes most similar to people in spacesuits are the most recent forms of Dogu, which are preceded by a long period of gradual evolution from very primitive and unimaginative clay representations of man. The Dogu were modeled from clay by ancient craftsmen of the advanced New Stone Age—these figures are 3 to 11 thousand years old. If, as is usual in archaeology, we construct an evolutionary series of statuettes from the earliest (the primitive) to the latest (the refined and fanciful) then we will see that all these hatches, filters, and other technical details of the spacesuits are in fact the result of a gradual artistic transformation, the stylization of earlier forms, originally quite realistically portraying parts of the human body.

Dealing with these features one by one the author notes that attention should be paid to one very important circumstance:

> All the ancient Japanese Dogu represent women. Evidently this is a manifestation of matriarchal principles in the ancient Ainu social structure, a result of the cult of the goddess of motherhood and fertility. Female breasts are distinctly visible on almost all the well-preserved statuettes. The breasts are usually bare and protrude, if only through an opening in the clothes.

Arutyunov concludes:

> I think it extremely unlikely that among the supposed visitors from the stars, not only should women decidedly predominate, but that they should pose before the ancient Ainu partially depressurized or, for that matter, having in any way unfastened their spacesuits.

Why It Sells: A Sociological View

The Japanese Dogu statuettes and fabricated Ecuadorian adventures further illustrate that von Däniken's books are not written to persuade the informed reader. They are a romanticist's fiction, and to examine or criticize the material at any greater length seems pointless.

Yet, it is tempting to dwell for a moment on the puzzle that the von Däniken books themselves present. If they are so bad, why are they so "good" (i.e., in terms of sales)? According to the British monthly *Encounter*, von Däniken's books tell us not about the ostensible subject matter but about the society which buys them so eagerly. By using the phrase "science fiction writer Erich von Däniken" *Time* signaled its readers that it didn't believe all this stuff but missed a central point, since most of von Däniken's readers obviously do not believe they are purchasing entertaining science fiction—nor does he so intend.

Von Däniken believes that one of the major reasons for the success of his books is religious uncertainty. Perhaps the astounding feats of the U.S. space program—men hovering in space, driving on the moon, and looking back to earth from beyond—basically disturb some people. For those who need certainty and stability in a world of jolting change and ambiguity, the attractions of von Däniken's books are that they are understandable and are presented vividly with firm conviction. They represent a substitution of faith for reason. Unfortunately, they can offer only a brief haven as they are so vulnerable to criticism. Still, many cults are alive today whose origins were equally shaky.

CHAPTER TWELVE

IS INTERSTELLAR TRAVEL POSSIBLE?

Tradition has it that Wan Hu was a local Chinese official of somewhat the same rank as Confucius had been at the start of his career. In any event, Wan Hu constructed a rocket-propelled flying machine whose exact appearance is confused by a welter of conflicting descriptions. The only points of agreement are that Wan Hu caused to be built a device consisting of a chair (or saddle), two kites, and forty-seven rockets. Coolies with torches closed in at a signal from Wan Hu, when he was seated, and lighted the rockets. There was a tremendous flash and Wan Hu was never seen again.

Carsbie C. Adams

JULES VERNE'S 1865 novel *De la Terre à la Lune* (*From the Earth to the Moon*) described a voyage to the moon by a three-man team, how they were propelled from a location in Florida (actually near Cape Canaveral) by a great gun, and how they were successfully picked up after splashing down in the Pacific Ocean off San Francisco. So uncannily accurate have been many of the science fiction prognosticators that one might be forgiven for supposing that what is fiction today becomes fact tomorrow. May we then turn to literary sources for instruction on the future of interstellar space travel?

Obviously not. For each bit of science fiction that has come true, a great deal more has proven false. We cannot know in advance which may prove reliable. Certainly we have no lack of accounts of men traveling to the realm of the stars. Not only do classical legends credit Greek gods with such powers of travel but current entertainment media also draw steadily on the same theme. Even the established term *astronaut*, deriving from the Greek word *astron*, meaning star, implies a traveler to the stars (although my dictionary cautiously limits itself to a "traveler in *interplanetary* space"). Hitherto, of course, our astronauts have limited themselves to cislunar space, the merest threshold to the vastness of interplanetary space, which itself is minute in extent compared to interstellar space.

Already unmanned space probes have traveled to Mercury, Venus, Mars, and Jupiter, and it is perfectly feasible to return probes to earth, as was first demonstrated in the case of the moon. Round-trip travel time to a planet (e.g., about two years to Jupiter) sounds feasible for astronauts, although not without problems. The first voyages around the world and the first expeditions to the Antarctic continent involved similar stays away from home under cramped and difficult conditions. We have no reason to doubt that manned spaceflight to the planets of our solar system is technically possible, if fiscally problematical.

Travel Time to the Stars

To answer the question of whether travel to the stars is possible we turn to experimental physics and observational astronomy—an approach that was successful in the design of the space vehicles launched to the moon and planets. One of the simplest bits of information to calculate is the *time that is needed for a round trip*, let us say to Proxima Centauri (which is often used as an example because of its closeness to the earth). First, we need to know the distance to this star, which is immense. We take a photograph of the sky showing the relationship of Proxima Centauri to the numerous faint stars in the surrounding star field. Another photograph is taken six months later. These two photographs will appear much the same, but because the earth has changed its position in space (being at the opposite side of its orbit) Proxima Centauri will appear slightly shifted with respect to its background. This process is called the *method of parallax* (the same principle is used in the range finder of a camera). The maximum displacement from the average position is the *heliocentric parallax*, a small angle that, in the case of Proxima Centauri, has been determined to be 1/270,000 of a radian (or 0.763 seconds of arc). This means that the distance to Proxima Centauri is 270,000 astronomical units. (An *astronomical unit* is the average distance from the earth to the sun [93 million miles] and provides a convenient measure for expressing the geometrical situation in interplanetary space.) Whereas light from our sun reaches the earth in 8 minutes, it takes 270,000 times longer to reach Proxima Centauri. We calculate $270{,}000 \times 8$ minutes $\approx 4\frac{1}{2}$ years; Proxima Centauri is therefore $4\frac{1}{2}$ light-years away. A space vehicle traveling at the speed of light on a round trip from earth to Proxima Centauri would take 9 years.

However, as has been well publicized since the time of Einstein, there is reason to think material objects cannot be propelled at the speed of light at all. Experience in the laboratory has consistently borne this out. We also know that the human body cannot safely be jerked instantaneously into motion but requires gradual acceleration. Thus, we need a flight plan whereby the spaceship accelerates gradually but steadily at a rate acceptable to the human body. The acceleration of a falling body due to gravity is suitable, since an astronaut accelerating through space at this rate, instead of feeling weightless, would be pressed down on the floor of his space vehicle so as to feel exactly the same weight as he does on earth. If we calculate how long it takes to approach the velocity of light, the answer is about a year, during which time half a light-year would be traversed. It will therefore take approximately 11 years for the round trip, if we plan to cover most of the distance traveling close to the speed of light following the flight plan of *Fig. 12*.

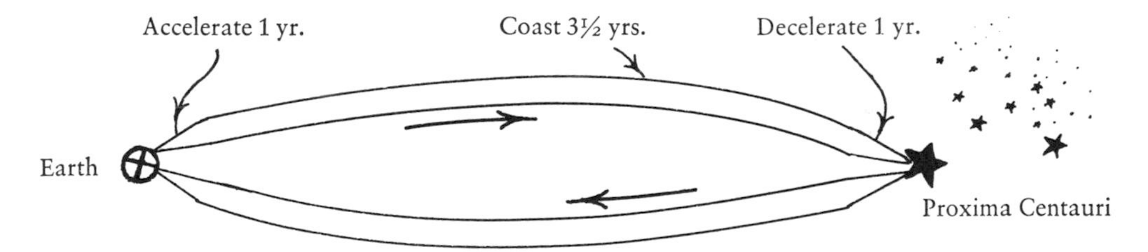

Fig. 12: Round-trip flight plan to Proxima Centauri occupying 11 years.

The Paradox of the Twins

There is also a bonus working in favor of the space traveler, often discussed under the title of "The Paradox of the Twins." Because of relativity effects, if a person departs on a space trip traveling at a velocity close to that of light, leaving his twin at home on earth, upon his return from space he will be younger than his twin. As seen from the earth, time on the traveler's spaceship will be slowed down. Yet, to the traveler, ship time flows at the normal rate. Such a situation is so far removed from everyday experience that it has been subjected to unusually critical scrutiny by theoretical physicists expert in the field of relativity. In the case of a 10-year trip such as we have described, only a year or two would be saved through the effects of this time dilation, but for long trips the "twin paradox" kindles the imagination. A 3,000-light-year trip, for example, which far exceeds the human life span, could, as far as people on board the spaceship were concerned, be completed in 30 years, according to some scientists. This tantalizing prospect has been assiduously entertained by writers of science fiction. (The following table shows a much more conservative, and I think more attainable, estimate of the effects of time dilation in space travel.)

Table 2: Duration of Space Flight Traveling at 99 Percent of Speed of Light

Distance	*Time for Space Crew*	*Time for Earth People*
10 light-years	2.8 years	20 years
100	28	202
1,000	277	2,020

Before leaving man-related considerations to delve into the rocketry required, we might note one uncomfortable factor. In interstellar space there are protons and electrons and occasional atoms which in themselves would not present a hazard, but if one tries to plow through them at nearly the speed of light they will bombard the spacecraft with great energy. Specks of interstellar dust will be even worse and will explode on impact with the vehicle. But for the purpose of discussion let us say that this problem can be taken care of by a sufficiently massive shield. Bear in mind that such a spaceship will not be small. As well as carrying provisions and life-sustaining equipment to last a decade, the hull itself will need to be rugged.

Mass Ratio Problems

To deliver a payload to its destination the launch vehicle must be very much heavier than the payload itself. (The takeoff mass exceeds the mass of the payload by a factor called the *mass ratio*.) The small module that splashes down in the ocean with the lunar astronauts aboard is the round-trip payload in the case of moon trips. When we compare this small module payload with the vast proportions of the multistory rocket that was needed to launch the payload, we see that the mass ratio is a large number.

With chemical fuels such as are now used, the outlook for space travel is not favorable. A calculation made by physicist Edward Purcell of Harvard puts interstellar travel in perspective. He assumes that the rocket will be propelled by controlled nuclear fusion. (Four molecules of hydrogen are converted to one of helium, by a thermonuclear reac-

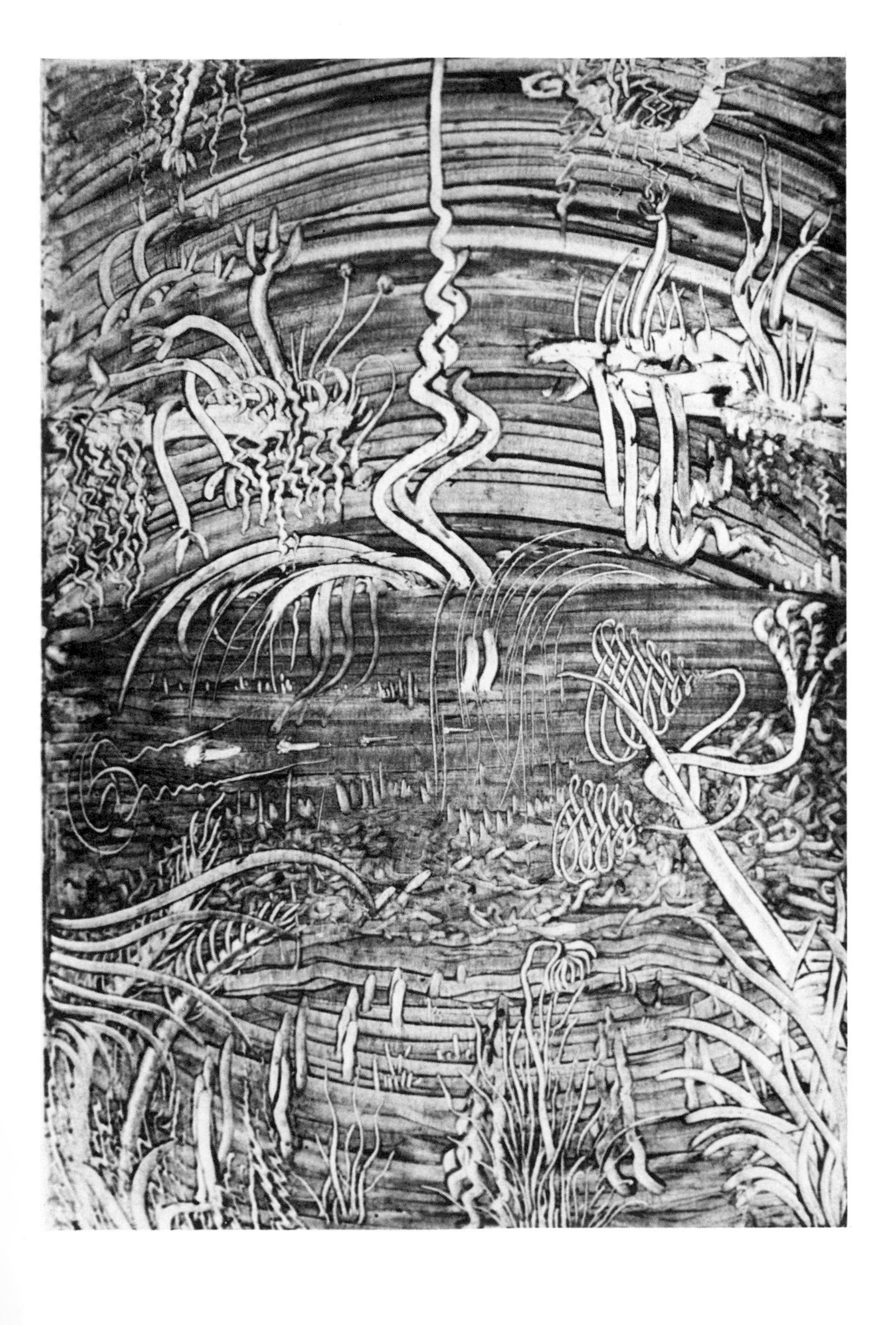

"All I know is when we got here this morning there they were."

Drawing by Handelsman; © 1961.
The New Yorker Magazine, Inc.

tion far more energetic than any chemical reaction, even more powerful than nuclear fission as currently used in nuclear power stations.) Although fusion occurs in hydrogen bomb explosions, two decades of stupendous efforts in several countries have failed to bring about *controlled* nuclear fusion. We cannot yet cause hydrogen nuclei to fuse into helium nuclei for use as a power source. Still, Purcell assumes that this will someday become possible. The amount of energy potentially available can be precisely calculated, and he assumes that 100 percent of this will be harnessed to expel the helium directly backward through the exhaust. Under these favorable conditions, the mass ratio required to accelerate a payload up to 99 percent of the speed of light and make a round trip to Proxima Centauri is one billion; the take-off weight would have to be *a billion times* the weight of the payload! As this is beyond the realm of feasibility, we must look further.

Annihilation of matter is the ultimate energy source under our present-day understanding of physics: it occurs in the laboratory when an electron coalesces with a positron to emit a flash of energy in the form of gamma rays (similar to X rays but having shorter wavelengths). The matter disappears completely. Although we do not know how to annihilate matter in bulk, we cannot be certain that energy production by the annihilation of matter will remain forever inaccessible. A rocket propelled by energy obtained from annihilation of matter is interesting to study since, whether or not it could ultimately be built, its theoretical performance permits us to set a *limit* to our expectations regarding space travel. Allowing for two accelerations and two decelerations to complete a round trip (as in *Fig. 12*) we find a mass ratio of 40 thousand; a 1,000-ton spaceship would have to start out with *40 million tons of fuel* to be annihilated for propulsion. The technical problems would be compounded by the fact that at takeoff the rocket would be radiating deadly, penetrating gamma rays equal in energy to the full solar power that falls on the earth. *Unless somehow screened, the gamma rays would exterminate life on earth.*

The basic difficulties encountered above, namely the hazard of plowing through the interstellar matter at high speed and the punishing mass ratios, arise from an impatient desire for a shortened voyage at high speed. The time schedule so imposed sets the acceleration required, and this in turn demands the massive rockets of at least 40 million tons.

A completely different approach has been taken by theoretical physicist Freeman Dyson of Princeton University. Neglecting entirely notions of haste, he concentrates instead on *efficiency* and *economics* and reaches the conclusion that in about 200 years interstellar voyages could begin. A payload of 10,000 tons, which he likens to a Noah's Ark, could reach the nearby stars, taking a few centuries to complete the trip. The cost, which would be prohibitive today, will in time be equivalent to that of the Saturn V rockets, each of which costs $100 million. The efficiency of the proposal, which requires a lift-off load of only 40,000 tons (a mass ratio of 4), is based on two bizarre designs, each of which utilizes repeated hydrogen bomb (fusion) explosions for propulsion. After allowing for all the obvious difficulties that this would entail, Dyson has established specifications that comply with the needs of the travelers in ways that can be costed and do not depend on the discovery of new technology such as controlled nuclear fusion or annihilation of matter. Purcell's and Dyson's studies are interesting demonstrations of feasibility. While their designs may never be implemented, we are left with the prospect that sometime in the next few centuries the price will be right to allow earthly communities to raft slowly out into interstellar space. Whether we will do this, or what may be the force of events impelling us to do so, cannot be foreseen.

CHAPTER THIRTEEN

THE ULTIMATE SCOPE OF TECHNOLOGY

The earth is the cradle of mankind, but one does not live in the cradle forever.

Konstantin Tsiolkovsky

As ONE LOOKS DOWN from an airplane it is impossible not to be struck by the man-made character of much of the scenery. Corn fields wave where there was once prairie, forests have vanished to make way for housing and more fields, and surface scars reveal that minerals are being taken from the ground. One can imagine progress continuing until the whole surface of the earth has been remodeled by the hand of man.

Further evidence of a one-way process is the increase in the number of humans with the passage of time. Although it is not known for certain what the population of the earth was in earlier times, this table indicates how world population has grown.

Clearly, the population is still increasing. If growth continues hyperbolically as it has with the annual *rate* rising as the population rises, there will be a catastrophe whose imminence is exhibited in *Fig. 13*. The population would become infinite in 2026 A.D. Of course, the popula-

Table 3: World Population Growth

Year	*Population* (millions)*	*Tons of Human Protoplasm*
1,000,000 B.C.	0.1	5,000
100,000 B.C.	1	50,000
10,000 B.C.	10	500,000
Christian era	100	5,000,000
1000 A.D.	200	10,000,000
1650	500	25,000,000
1750	700	35,000,000
1850	1,000	50,000,000
1950	2,500	125,000,000
1975	4,000	200,000,000

* Rounded figures based on Edward S. Deevey, Jr., "The Human Population," *Scientific American*, vol. 203, No. 3, Sept. 1960, pp. 194–204; plus *Population Bulletin*, vol. 18, No. 1, 1962.

tion cannot become infinite; growth must slacken before then. But unfortunately, starvation on an unprecedented scale, in at least one major country, may be necessary to get the active attention of governments and people outside that area, if the restraint of growth is to be voluntary.

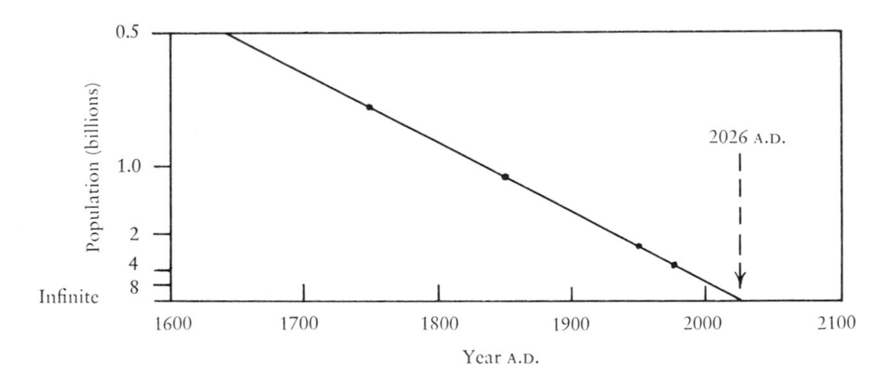

Fig. 13: Showing how the world population, if projected into the future, reaches infinity in 2026 A.D. (from Sebastian von Hoerner, in E. Scheibe and G. Süssmann, eds., "Population Explosion and Intersellar Expansion" (*Einheit und Vielheit*, Vandenhoeck and Ruprecht, Göttingen, 1972).

Either mankind will stabilize his numbers of his own volition or his increase will be checked catastrophically, country by country. But whether we reach a population limit in an orderly way or not, a more or less definite limit presumably exists. It is determined ultimately by the earth's supply of energy, which controls the amount of food that can be grown, harvested, and distributed.

The earth as a whole is becoming a single interconnected system as a result of improvements in communication and transportation. In the past, rising growth curves in isolated regions have collided drastically with limits set by food supply. Examples occurred in this century in Russia (1918-22, 1932-34), China (1920-21), and Bengal (1943). In each of these cases deaths by famine were in the millions, but less drastic isolated instances occur somewhere every year. Bulk transportation of food has reduced the isolation of populations. But, even now, shipments only mitigate the effects of food shortages and represent no solution at all to the impending *global* crisis.

Although we have just presumed a ceiling to the total population of the earth, historical examples of insufficient food have *not* led to ceilings being established; instead, setbacks or pauses in the expansion in localities affected have occurred. Various explanations are possible: clearing new land, introducing or improving fertilizers, breeding more productive strains, inventing pesticides, and so on. In earlier times, innovations that permitted population breakthroughs included the transition from food gathering to big game hunting, the introduction of agriculture and animal husbandry, and the industrial revolution.

But surely the global ceiling will prove to be the ultimate limit to population growth, since past solutions will not be available when the whole planet is occupied and the best methods for food production have been found. Still, man is ingenious. Could the next breakthrough, bringing salvation in the nick of time, come in the form of spaceflight? The answer to this is an unequivocal *no*! At our present annual 2 percent growth rate, even if we could ferry people to the nearest 30,000 habitable planets, they would all be occupied in only 500 years. After that, even if the volume occupied could be expanded at the speed of light, the annual increase in volume would be below the needed 2 percent. In practice, in the few decades left, we will not have the resources to feed our people, let alone lift them into space. Zero population growth is the only way to go.

This is not to say that once the immediate crisis is weathered man will not expand into the solar system; Freeman Dyson has speculated on the new limits that would be imposed for a civilization that had learned to harness the energy production of its star. (For us the total solar power output of 4×10^{23} kilowatts would become the new limit.) The energy limit does not in itself tell us to what extent the population might expand. A second limiting condition appears now, namely the amount of *matter* available. A civilization that could harness the energy production of its star, Dyson says, might want to populate the largest possible area, using for these goals whatever matter is available. For us, this would mean taking apart the largest planets (Jupiter and Saturn) and building from this matter a spherical shell around the sun somewhat larger than the earth's orbit. Dyson does not advocate disassembling the earth itself.

The third column in *Table 3* shows the tonnage of human protoplasm corresponding to each population entry, calculated by allowing the rounded figure of 100 pounds per person. Expressing the population increase in this blunt and unusual way raises the question: Where is all this meat coming from? A century ago, when there were so many tons less of us, where were the atoms and molecules to be found that today are a part of people? Clearly what is today human protoplasm was then dispersed through the land, sea, and air in one form or another. The air we breathe, our food, and our drink are progressively being converted to flesh and blood. At the present time this withdrawal of inanimate matter from the earth and its atmosphere amounts to only an infinitesimal trifle, but in the long run we cannot go beyond what the available matter allows. The earth limits us to a total of 6×10^{21} tons. If we include the matter of the major planets but do not count the hydrogen and helium, we have about 30 times more than this amount.

Dyson has shown how it is theoretically possible to use solar energy to disassemble a planet and rearrange the pieces so as to intercept the maximum solar energy. As he says, "There is nothing so big nor so crazy that one out of a million technological societies may not feel itself driven to do, provided it is physically possible."* If Jupiter were to be taken to pieces and redistributed in orbits in space within a shell about the same distance from the sun as the earth, the sun's energy could then be caught (save for a fraction filtering through chinks) and minerals and other materials of Jupiter's interior would be available for use. The length of time required to restructure the entire planetary system into solar collectors would be around 100,000 years.

Any astronomers on another planet would notice a striking transition in the appearance of our sun if this project proceeded. The sun as they had known it would be mostly obscured and would begin to appear as a much larger, much darker object. By the law of conservation of energy, just as many kilowatts would flow out of this dark object as used to flow from the sun when it was visible. The difference is that the energy would be mainly in the form of *infrared heat radiation instead of visible light*. Dyson proposes that the sky be searched for infrared objects on the chance that some big technology may have developed astroengineering on this scale, thereby creating a distinctive infrared object in our night sky. He states, "One would look in particular for

* Freeman Dyson, "The Search for Extraterrestrial Technology," *Perspectives in Modern Physics*.

irregular light variations due to starlight shining through chinks in the curtain, and for stray electromagnetic fields and radio noise produced by large-scale electrical operations (not necessarily carrying any 'message')." No such infrared objects emitting faint starlight have been found; on the other hand infrared astronomy is a young subject, and we are still far from having surveyed all the sources with a power output comparable to our sun's.

Fascinating engineering can be indulged in as one tries to work out the best ways to mine a planet and launch the pieces into orbit, but the details will depend on taste. The essence of this discussion is that we may envisage the end product before elaborating the means of achieving it!

I find no scientific flaw that would negate Dyson's concept. There *are* technical difficulties with the shell, whose component parts will be in relative motion as they move in accordance with the law of gravity, but there is no basic objection that I am aware of that would prevent a fairly close approach to total usage of the available mass and energy.

However, merely because something is possible does not mean that it will inevitably occur. As time passes and further imaginative studies are made in astroengineering, alternative courses of action will be offered to our descendants for the furtherance of their purposes, whatever they may be. For example, we could investigate the cost of disassembling Ganymede and Callisto (satellites of Jupiter) or Titan (satellite of Saturn), all three bigger than our moon, to see if the cost is favorable. It would prove economical to quarry satellites or asteroids before Jupiter. When Jupiter and the remaining planets have been digested, we are face-to-face with Dyson again. We will have utilized the ultimate resources of our culture, the sun's power output and the planetary system's mass—as long as we remain with our present sun. But even if *this ultimate ceiling* were to be reached, would it prove to be merely another pause in the growth of the human race to be followed by expansion to the stars?

CHAPTER FOURTEEN

THE COLONIZATION OF SPACE

Man will not ultimately be content to be parasitic on the stars but will invade them and organize them for his own purposes.

J.D. Bernal in *The World, the Flesh, and the Devil*

ALTHOUGH IT IS NOT POSSIBLE to foresee what the specific motivation might be, earthly life may someday diffuse into space. With the pressures of crowding and its attendant pollution, a part of society may consider space to be a better habitat than earth. Or, as suggested by J.D. Bernal, the driving force for interplanetary colonization may turn out to be competition for sunlight and meteoric matter.

When the colonists depart, what will their new home be? One might think first of a giant structure resembling Skylab. But, as self-sufficiency is a goal for a sustained colony in space, we should keep in mind long-term requirements of self-support: agriculture, animal husbandry, a method for intercepting the necessary solar energy. Such a home in space must be large, but can be smaller than one might initially think. Bernal described space colonies of 20,000 to 30,000 inhabitants living in the interior of globes 10 miles in diameter.

O'Neill's Space Colonies

A proposal for the colonization of space has been worked out by Gerard K. O'Neill of Princeton University. He envisages a cylindrical vessel in space, 4 miles in diameter and 16 miles long. With air inside, it spins on its axis once in 1.9 minutes producing a centrifugal force exactly equivalent to gravity on earth. A person could stand up inside the cylinder, and feel that he was walking on level ground. However, his visual impression would be of standing at the bottom of a rounded valley. Trees growing on the sides of the valley would give the impression of leaning downhill (they would have grown up toward the axis, in the direction opposite to the centrifugal force). To brighten the habitat, O'Neill divides the curved surface of the cylinder into six long strips, each two miles wide, and makes three of them transparent windows, thus letting sunlight through onto the three solid strips of steel and soil which will form the inhabited valleys.

Because of the scattering of sunlight by the air molecules contained within the cylinder, the sky will appear blue, giving a feeling of naturalness to the habitat. Since the whole cylinder is rotating once each 1.9 minutes, the sun would, however, be winking on and off in a most unnatural way as the three windows pass successively around. To keep the sun fixed in the habitat's sky we need three giant mirrors to deflect the sunlight through the windows, at the same time arranging to point the axis of the cylinder directly at the sun. To appreciate this arrangement consult the graphic depiction in *Fig. 14*.

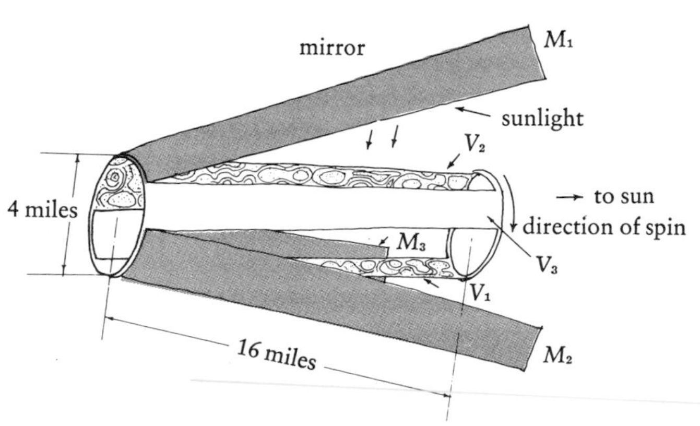

Fig. 14. O'Neill's design for space colony, showing the arrangement of three valleys (V_1, V_2, V_3,) separated by windows. Valley V_1 is illuminated by sunlight reflected from mirror M_1, and so on.

Fig. 15 View along one of the valleys of O'Neill's cylindrical world, showing lakes, villages, hills, and afforestation projects. The other two valleys are visible, appearing as if suspended between windows of blue sky.

If desired, sunrise and sunset, as well as seasons, can be simulated by tilting the mirrors. One of O'Neill's purposes is to provide an attractive habitat—one perhaps preferable to what might be available on earth in years to come. In pursuit of this ideal, the valleys will be laid out as meadows, forests, lakes, villages, and mountains—even to the extent of copying areas of the earth that are celebrated for their scenic beauty (see *Fig. 15*). The ends of the cylinder will be sealed to keep the air in through the use of simulated mountains up to two miles high. They will be strange mountains to climb. The lack of oxygen noticed on two-mile-high mountains on earth will not be felt, and gravity will diminish rapidly as one climbs until a state of virtual weightlessness is at last reached. A vacation of mountain climbing would be a fascinating experience.

The first community would be built mainly from lunar surface materials, with only about 2 percent of the mass brought up from earth. Hydrogen from the earth, combined with oxygen from the lunar surface oxides, would provide water at a big saving in transport requirements. All the matter would be recycled. The later, bigger communities would obtain their materials from the asteroid belt—rich in hydrogen, carbon, and nitrogen as well as in the metals, minerals, and oxygen that are common on the moon.

Many details have been considered by O'Neill that must be omitted here, but may be consulted in the publication *Physics Today*, Sept. 1974. Interesting problems of structure and orbital mechanics arise. One of these is to prevent the spinning cylinder from gyrating like a top; to achieve this O'Neill counteracts the spin by building the cylinders in oppositely rotating pairs. Many biological decisions arise that will be intriguing to explore. How will the balance of nature be simulated and maintained? In O'Neill's plan, for example, birds will be introduced, but not flies and mosquitoes. What will the birds feed on, plants or insects? If insects, what will the insects feed on? Possibly an artificial ecology can be designed, but I wonder whether it will remain under control or develop a direction of its own.

So different are conditions in space from what we are accustomed to on earth that it will come as a surprise to learn how the inhabitants of these two adjacent cylinders can visit each other. Calculation shows that a cylinder 4 miles in diameter rotating once in 1.9 minutes has a tangential speed of 400 miles-per-hour. A space shuttle parked on the outside of the cylinder would have to be tied down in order not to be flung off at this speed. However, if released while pointing in the right direction, the shuttle would take off at 400 miles-per-hour and could dock on the outer surface of its parallel and oppositely rotating cylinder nearby, touching down gently at the moment of arrival when the shuttle velocity would exactly equal that of the cylinder. No rocket engine would be required, and no fuel, just a good docking system and good timing (see *Fig. 16*).

Weather would result from the draft of warm air rising from the sun lit valley bottoms, carrying moisture with it. Perhaps the moisture would condense into clouds or fog. Would rain fall only at night, and would the clouds form over the valleys? Would there be circulating winds? It is difficult to say what would happen in this system if left to

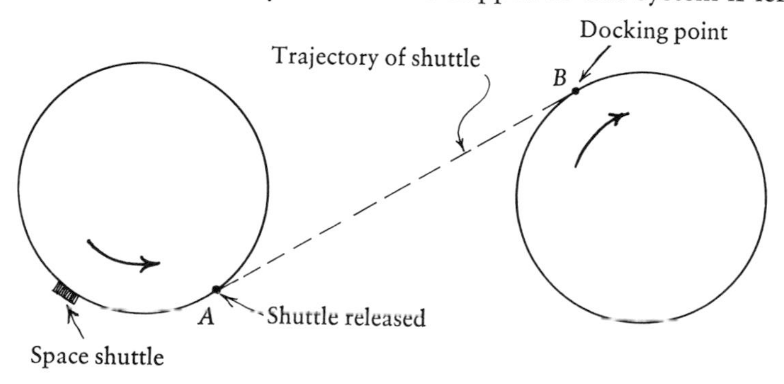

Fig. 16. A space shuttle, released at *A*, coasts freely without using fuel to docking point *B*, where it makes a perfect landing.

itself. Perhaps the water would tend to condense on the cold windows where it would be pressed into place by centrifugal force, forming lakes which would be used as sources of irrigation water. If so, no rain need ever fall. If rain was desired, perhaps small modifications to the cylinder (making it more like a hexagonal prism, giving it a slight waist like an hour glass, or varying the sunlight by tilting the mirrors) might make it possible to choose and control the climate. Should cylinders continue to be constructed, the experience gained might lead to a diversity of habitats.

Although imitation of terrestrial environments may be a good guiding principle, a cylinder with its broadside to the sun might be a perfectly satisfactory habitat, in which case the giant mirrors could be dispensed with. No dark shadows are to be expected within the cylinder even though the sun apparently races around outside—when any one valley is in shadow there will generally be some light reflected from sunlit strips of the other valleys just a few miles away. In addition, scattered light could be provided by making the windows of a diffusing substance. (Condensation on the interior of the windows would have the same softening effect on the shadows.) The net result could be similar to a day experienced on the North Atlantic coast when low patchy clouds racing overhead maintain a continual interplay of light and shade. While this is not the same as Mediterranean sunlight, it is pleasant to many people and is certainly natural to the earth. Another natural phenomenon is the long twilight of subarctic regions which, while confusing to visitors from temperate zones, is perfectly natural to the natives. In fact, man is so adaptable that great freedom is possible in arranging suitable habitats for him in space, austere as well as utopian. No doubt many ingenious designs will be pursued in coming years.

Space Industries?

After this new miniworld becomes self-supporting, how will it repay the loan that enabled it to be assembled with the aid of earth resources in the first place? It is conceivable that the inhabitants could finance their launching from their own capital, but it is hard to see how the earth could afford to spawn new offspring indefinitely. I feel that the urge to expand will lead, by one route or another, to continued colonization of space and utilization of its mass and energy resources. If a way could be formulated whereby the newly launched company could amortize its obligation to earth, I would expect it to be followed.

Since space is a high vacuum useful for special assembly techniques, one might think of computer manufacture as a suitable industry in space. Computer manufacture requires high-vacuum conditions, and

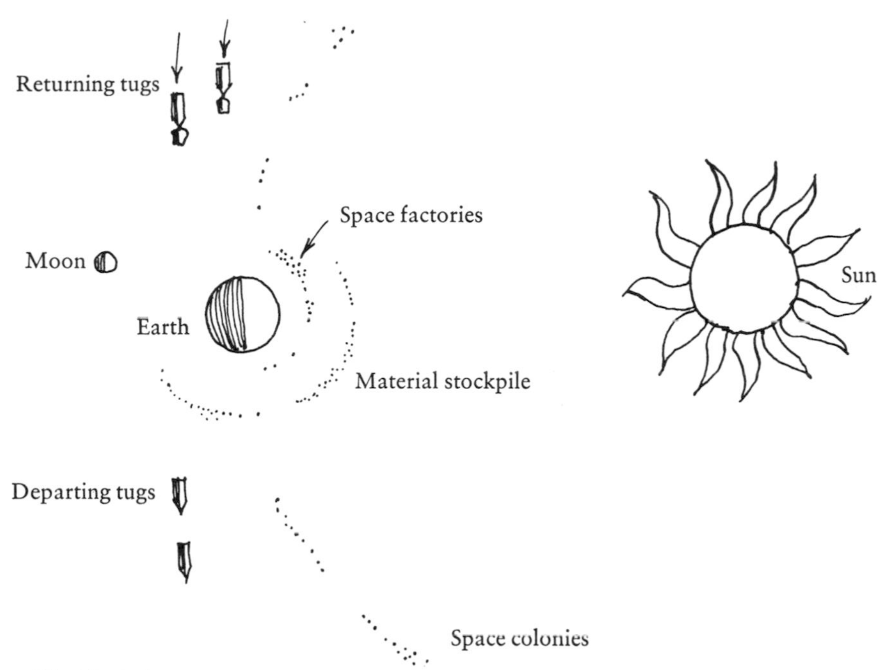

Fig. 17. Colonies are shown spreading out along the earth's orbit. Tugs are departing for the asteroid belt and others are returning, pushing asteroids before them.

the computer market will be expanding for a long time. Tourism is another possible space industry. However, working against any commerce between the earth and space colonies will always be the heavy price of lifting or lowering material through the earth's gravitational field. Therefore, saleable services not involving transfer of mass to or from the earth's surface should be sought first.

The two candidates that occur to me are manufacture of space structures and the supply of power. All sorts of space vehicles and space structures will be needed as the activity of space colonization gathers momentum: space tugs for going to the asteroid belt and back, bringing iron and stone for the space factories; space shuttles for the trips to earth; the shipyards, the local vehicles, the town and its supporting services (see *Fig. 17*). The development of these facilities will take many years. But in time, as space mining delivers more and more raw material to the construction sites in space, dependence on earth material with its costly freight charges can be relieved.

Given the capital investment in means of production, the space community can contribute an increasing share to the earth-originated venture. In the long run, the New Space World may exceed the earth in population and resources, just as on earth the Americas, after several hundred years of intense colonization, have caught up with and exceed, in many capacities, the communities that colonized them. However, it may be noted that shipbuilding, which received a tremendous stimulus from the colonization of the New World and the advent of world trade, has remained a major European activity over the half-millennium that has elapsed since the days of Prince Henry the Navigator. I envisage an enduring phase in which space ventures remain a major earthly activity, while a growing space community makes welcome contributions. In a way, this may be likened to the American contribution to world shipping through the design and building of the clipper ships of 1833-69. However, the parallel would be closer if building materials in Europe had lain at the bottom of a hole hundreds of miles deep. The cost of transporting raw materials through space compared with that of lifting materials from earth favors an economy based on the mining of space.

While the power for manufactures in space should come directly from the sun, very large amounts of power will also be required on earth. How could the space community help with this? What the relative costs of nuclear fusion energy and solar energy will be is not known. But if solar energy is a substantial component of the terrestrial energy budget there will be one awkward feature: an extensive geographic area has to be covered with solar energy collectors. A given collecting area assembled in space would free the earth's surface for other use; the energy collected could be radiated to earth in a condensed form that could be received on a much smaller collector.

With power handling and space manufacturing as an economic base, the expansion of space colonies will be linked directly to the mining of materials in interplanetary space. In addition to the moon and asteroids there will be the numerous satellites of Jupiter and Saturn, and ultimately the planets themselves. *We live now on a satellite adrift in space —why would man not make more satellites to his taste if it came within his means?*

Should this colonization program proceed in the direction indicated here, a great fleet of worlds will eventually accompany the earth in its orbit around the sun. At the very basis of this development is the possibility of obtaining more of the sun's energy and using more of the matter of the solar system. It is the same vision that Tsiolkovsky, the great Russian space pioneer, described in his book *Dreams of the Earth*, published in 1895. It is a step toward Dyson's world in which a shell ultimately envelops the sun to capture all its energy.

CHAPTER FIFTEEN

HEIRS TO A GALACTIC CULTURE?

Empty space is like a kingdom, and heaven and earth no more than a single individual person in that kingdom. Upon one tree are many fruits, and in one kingdom many people. How unreasonable it would be to suppose that, besides the heavens and earth which we can see, there are no other heavens and no other earths?

Teng Mu, 13th century philosopher

HAVING EXPLORED the current theories about extraterrestrial life presented in this book, you must now judge the probabilities for yourself. Are we alone? Do we have neighbors? Will we make contact? My opinion is that we are not alone—that we are not the only community to have gained a knowledge of the laws of nature and to have begun exercising control over nature. Somewhere in this galaxy or another I think there is other intelligent life. Where they are and when and whether they will reveal themselves I do not know, nor can I propose an immediate action on our part that would tell us.

My recommendation is to continue to gain an improved appreciation of the overall problem of contact and communication. Modest though our present understanding is of the restrictions imposed by the immensity of space and the eons of time, we now have a much clearer perspective than did the scientists in the 1950s and before. Further study will sharpen our knowledge and prepare us to deal with the arrival of a first message or to take the initiative ourselves through some course of direct action.

When the opportunity for contact arises it will be costly, in both money and time. Yet we will seize the opportunity, for reasons that will be compelling at that time. I do not know *what* the reasons will be, but here are suggestions. Peace may continue to elude human grasp, despite sincere effort, to the point where information about the social organization of an outside planet will be highly valued. Radioactive waste may continue to accumulate and the concerted effort to generate clean power by hydrogen fusion may limp along unsuccessfully, as it has since World War II, until even a small hint about the appropriate physics from an outsider will be urgently desired. Plain curiosity has often been the motivating force for large enterprises by private parties, but is not usually acceptable per se to government bodies administering sizable public funds. Nonetheless, curiosity is a strong force that triumphs in various guises, often by placing emphasis on byproducts valued by the payer. There will always be non-agnostics, and future discoveries about the physical universe, including the detection of other life, may furnish such persons with compelling arguments that exploration for intelligent life would answer deep questions of a religious nature about the meaning of life.

Not only would *we* need to justify a costly diversion of resources to a search for intelligent extraterrestrial life, but so would *they*. In this book we encountered serious proposals envisaging the sending of probes or powerful radio signals toward the earth, but we gave no attention to *why* a responsible extraterrestrial agency would make any sizable expenditure when no return would be forthcoming for centuries, if then. *Much thought should be given to this question because their motivation would affect both the nature of their effort to contact us and our strategy for facilitating contact.* The several justifications suggested above should be studied, and there would be others. For example, a delay of centuries sounds unacceptable to us because it exceeds our life span; but on another planet a century might be only a fraction of the lifetime of an individual. Also, although the cost of interstellar contact looks high to us, further study may suggest a relatively inexpensive method.

One way or another, I believe that an opportunity to contact extraterrestrial intelligence, should it occur, will be grasped, and that humanity will enter a new phase of evolution. We have passed through prebiological chemical evolution, Darwinian evolution culminating in the expansion of the brain to its present size and quality, and cultural evolution starting from the time when accumulated tradition came to be transmitted by word of mouth. Contact with extraterrestrial life would render us heirs to galactic culture. Would it set us on a new and higher path?

READER'S GUIDE

Chapter 1: Are We Alone?

Books on the general subject of intelligent life in space include *We Are Not Alone* by Walter Sullivan (New York: McGraw-Hill, 1964), *Intelligent Life in Space* by Frank D. Drake (New York: Macmillan, 1967), and *The Cosmic Connection* by Carl Sagan (New York: Doubleday, 1973), all of which are written at a popular level and are enjoyable reading. Broad coverage at a higher level is supplied in *Intelligent Life in the Universe* by I.S. Shklovsky and Carl Sagan (San Francisco: Holden-Day, 1966) which gathers together considerable detail on all branches of science that bear on intelligent life. *Interstellar Communication* ed. by A.G.W. Cameron (New York: Benjamin, 1963) is an indispensable anthology* of the primary sources on which much subsequent writing has been based; sprinkled among the numerous readable papers are some technical ones, but Cameron's collection is especially valuable in clarifying what the originators of the ideas actually said. For a sequel see *Interstellar Communication* ed. by A.G.W. Cameron and Cyril Ponnamperuma (New York: Benjamin, 1974). The present author's principal contributions to the subject can be found in the two Benjamin books. *Communication with Extraterrestrial Intelligence* (Cambridge: MIT Press, 1973) ed. by Carl Sagan presents the work of 54 contributors to a conference held in Byurakan under the auspices of the Academies of Science of the U.S. and U.S.S.R.

Other readable books are *Habitable Planets for Man* by S.H. Dole (New York: Blaisdell, 1970), *Intelligent Life in Space* by Frank D. Drake (New York: Macmillan, 1967), and *Life: Its Nature, Origin, and Development* by A.I. Oparin (London: Oliver and Boyd, 1961).

For general astronomical background, a wide range of popular books on astronomy is available. For more depth, one may choose among many elementary astronomy texts. The following books, which are mostly non-mathematical and cater largely to humanists, are especially recommended: *A Survey of the Universe* by Donald L. Menzel, Fred L. Whipple, and Gérard Vaucouleurs (New York: Prentice-Hall, 1970); *New Horizons in Astronomy* by John C. Brandt and Stephen P. Maran (San Francisco: W.H. Freeman, 1972); *Exploring the Cosmos* by Louis Berman (Boston: Little, Brown, 1973); and *Introductory Astronomy and Astrophysics* by Elske v.P. Smith and Kenneth C. Jacobs (Philadelphia: W.B. Saunders, 1973). Books about the solar system are given under Chapter 2.

Chapter 2: Velikovskian Vermin

The basic references are the Book of Exodus and Immanuel Velikovsky's three books, *Worlds in Collision* (New York: Doubleday, 1950), *Ages in Chaos* (New York: Doubleday, 1952), and *Earth in Upheaval* (New York: Doubleday, 1955).

An exceptionally well-illustrated layman's introduction to planetary astronomy is

* Items marked with an asterisk can be found in this anthology.

Planets by Carl Sagan and J. Leonard (New York: Life Science Library, 1966); another good background book on related topics is *Meteors, Comets and Meteorites* by G.S. Hawkins (New York: McGraw-Hill, 1964). For more about meteorites see *Out of the Sky* by H.H. Nininger (New York: Dover, 1952) and *Find a Falling Star* (New York: Paul S. Eriksson, 1972).

For description and evaluation of contending theories on the advancing frontiers of science see *The Origin of the Solar System* edited by T. Page and L.W. Page (New York: Mullin, 1966); *The Origin of the Solar System* by H.P. Berlage (New York: Pergamon, 1968); "The Origin of the Solar System" by D. ter Haar in *Annual Reviews of Astronomy and Astrophysics*, vol. 5, p. 267, 1967; and *Adventures in Earth History* by P. Cloud (San Francisco: W.H. Freeman, 1970). "History of the Lunar Orbit" by P. Goldreich in *Reviews of Geophysics*, vol. 4, p. 411, 1966, is a remarkable chapter in the theory of celestial mechanics.

The exciting story of continental drift revealing how the original single landmass of Pangaea broke into Laurasia and Gondwanaland, which in turn split apart into the continents we now know, is assimilated in a well-illustrated and readable anthology of articles from *Scientific American* entitled *Continents Adrift* edited by J.T. Wilson (San Francisco: W.H. Freeman, 1971). The story is based on seismological and sea-floor studies that are detailed in *Plate Tectonics and Geomagnetic Reversals* by Allan Cox (San Francisco: W.H. Freeman, 1973).

Chapter 3: Is Man Unique?

For an article refuting the concept of man's preeminence, see R.N. Bracewell, "Life in the Galaxy" in *A Journey Through Space and the Atom* edited by S.T. Butler and H. Messel (Sydney: Nuclear Research Foundation, 1962).* The principle that we are merely mediocre argues against the assumptions that the closest likely star is inhabited and that *our* first contact will be the first contact ever made, as explained in R.N. Bracewell, "Communications from Superior Galactic Communities," *Nature*, vol. 186, p. 670, 1960.* Further ideas of interest on this subject are found in "The Search for Signals from Other Civilizations" by Sebastian von Hoerner, *Science*, vol. 134, p. 1839, 1961.* A chapter entitled "The Assumption of Mediocrity" by I.S. Shklovsky and Carl Sagan in *Intelligent Life in the Universe* (San Francisco: Holden-Day, 1966) also develops this theme. For the sobering thoughts about the possibility that man in unique, see W.H. McCrea, "Astronomer's Luck," *Quarterly Journal of the Royal Astronomical Society*, vol. 13, p. 506, 1972.

Chapter 4: Should We Talk to Them? And How?

Radio telescopes are photogenic instruments which have revolutionized astronomy; the story behind them has been told from different viewpoints. Science writer J. Pfeiffer provides the anecdotal *The Changing Universe* (New York: Random House, 1956). Sir Bernard Lovell mixes the human side with the astronomy in *The Story of Jodrell Bank* (Oxford: Oxford University Press, 1968) and *Out of the Zenith: Jodrell Bank 1959-1970* (New York: Harper and Row, 1973). Radio-astronomy pioneer J.S. Hey illuminates the cast of characters and the interwoven plots in *The Evolution of Radio Astronomy* (New York: Science History Publications, 1973).

The principles of various kinds of radio telescopes are given in "Radio Astronomy Techniques" by R.N. Bracewell in vol. 54 of *Encyclopedia of Physics* edited by S. Flügge (Berlin: Springer, 1962) and in *Radio Telescopes* by W.N. Christiansen and J.A. Högbom (Cambridge: Cambridge University Press, 1969).

Chapter 5: Project Cyclops

Project Cyclops, A Design Study of a System for Detecting Extraterrestrial Intelligent Life ed. by Bernard M. Oliver (NASA Report No. CR 114445, 1973) can be obtained from Dr. John Billingham, NASA/Ames Research Center, Code LT, Moffett

Field, California 94035. The Cyclops report is not only a design study, but is also a compendium of up-to-date information and a source of technical references.

Chapter 6: Making Radio Contact

The stimulus to scientific thought on contact with extraterrestrial life via radio was "Searching for Interstellar Communications" by G. Cocconi and P. Morrison, *Nature*, vol. 184, p. 844, 1959.* Project Ozma was Dr. F.D. Drake's reaction to this stimulus. His writings, "How Can We Detect Radio Transmissions from Distant Planetary Systems?" *Sky and Telescope*, vol. 19, p. 140, 1959,* "Project Ozma," *Physics Today*, vol. 14, p. 140, 1961, and *Intelligent Life in Space* (New York: Macmillan, 1966) are interesting descriptions. The Cyclops report (see Chapter 5) brings the subject of radio contact up to date.

Chapter 7: Intelligent Neighbors: How Far, How Many?

The idea of average longevity of a civilization was introduced in "Communications from Superior Galactic Communities" by R.N. Bracewell, *Nature*, vol. 186, p. 670, 1960,* along with the notion that durability or quasi-permanence might be achieved as a consequence of interstellar contact. Examples of the technical use of this concept are found in "The Search for Signals from Other Civilizations" by S. von Hoerner, *Science*, vol. 134, p. 1839, 1961,* and in subsequent studies such as Shklovsky's and Sagan's *Intelligent Life in the Universe* and the Cyclops report. Theoretical refinements will be found in "A Formulation for the Number of Communicative Civilizations in the Galaxy," by J.G. Kreifeldt, *Icarus*, vol. 14, p. 419, 1971. An estimate of other methods for estimating the abundance of intelligent life are suggested in *Of Stars and Men* by H. Shapley (Boston: Beacon Press, 1958).

Chapter 8: Interstellar Messengers

Excluding numerous instances in science fiction, the published discussions of messenger probes are found in the following sources: *Vselennaia Zhizn' Razum (Universe, Life, Intelligence)* by I.S. Shklovsky (Moscow: Soviet Academy of Sciences, 1962) and *Intelligent Life in the Universe* by I.S. Shklovsky and Carl Sagan (San Francisco: Holden-Day, 1956). The author's papers on messenger probes are "Communications from Superior Galactic Communities," *Nature*, vol. 186, p. 670, 1960;* "Life in the Galaxy" in *A Journey Through Space and the Atom*, ed. by S.T. Butler and H. Messel (Sydney: Nuclear Research Foundation, 1962);* and "Interstellar Probes" in *Interstellar Communication* ed. by A.G.W. Cameron and Cyril Ponnamperuma (New York: Benjamin, 1974). Duncan Lunan explains his story about a probe from Epsilon Boötis in *Man and the Stars* (London: Souvenir Press, 1974); the story depends on an interesting code that receives close scrutiny in "The Opening Message from an Extraterrestrial Probe" by R.N. Bracewell, *Astronautics and Aeronautics*, vol. 11, p. 58, 1973. For an explanation of techniques used in determining the age of meteorites and the damage they suffer in space see "Meteorites and Cosmic Radiation" by I.R. Cameron, *Scientific American*, p. 65, July 1973.

Chapter 11: The Chariots of von Däniken

For fascinating reading on the variety of technical skills of ancient Egypt (including the transport of tremendous loads) see *The History of Technology* edited by C. Singer, E.J. Holmyard and A.R. Hall, Vol. 1 (New York: Oxford, 1954).

Chapter 12: Is Interstellar Travel Possible?

An influential paper, "Radioastronomy and Communication through Space," by E. Purcell, U.S. Atomic Energy Commission Report BNL-658, 1960* demonstrates that round trips to stars by humans within a human lifetime is preposterous. Further technical information is given in "The General Limits of Space Travel" by S. von Hoerner, *Science*, vol. 137, p. 18, 1962.* The scale of engineering and the economic aspects of the problem are indicated in "Interstellar Transport" by F.J. Dyson, *Physics Today*, vol. 21, p. 41, 1968. The physics of time dilation is explained by Bernard M. Oliver in "The View from the Starship Bridge and Other Observations," *IEEE Spectrum*, September 1974; for the application to space travel see Carl Sagan, "Direct Contact Among Galactic Civilizations by Relativistic Interstellar Spaceflight," *Planetary and Space Science*, vol. 11, p. 485, 1963.

Chapters 13 & 14: The Ultimate Scope of Technology & The Colonization of Space

The basic source of population statistics is *Demographic Yearbook* published annually by the United Nations; see also the periodical *Population Bulletin* and E.S. Deevey, "The Human Population," *Scientific American*, September 1960. The entire September 1974 issue of *Scientific American* is devoted to articles on human population and suggests further reading. For factors that will limit growth see "Population Explosion and Interstellar Expansion," S. von Hoerner, in *Einheit and Vielheit*, ed. by E. Scheibe and G. Süssman (Göttingen: Vandenhoeck and Ruprecht, 1972).

Space structures for catching the sun's energy are described in "Search for Artificial Stellar Sources of Infrared Radiation" by Freeman J. Dyson, *Science*, vol. 131, p. 1667, 1959, and in "The Search for Extra-Terrestrial Technology" by Freeman J. Dyson in *Perspectives in Modern Physics* (New York: Interscience Publishers, 1966).

For more on space colonies see "The Colonization of Space" by Gerard K. O'Neill, *Physics Today*, vol. 27, no. 9, 1974. A less detailed plan for space colonies was presented in 1929 in *The World, the Flesh, and the Devil* by J.D. Bernal (Bloomington: University of Indiana Press, 2nd ed. 1969) along with other interesting thoughts such as achieving immortality through interconnected brains.

INDEX

ABOUT THE AUTHOR

According to a current collection of student reports on instructors, "Bracewell is truly a weird professor." This designer of radio telescopes spends his spare time studying trees and importing seeds of rare plants which might enjoy the local climate. He is compiling a book covering all the trees to be found on the Stanford campus.

A Professor of Electrical Engineering, Ronald Bracewell is a native of Australia where he received degrees in mathematics, physics, and electrical engineering. After four years in the field of microwave radar, he went to the Cavendish Laboratory of Cambridge University, obtaining a PhD in physics in 1951. He lectured in the Astronomy Department of the University of California, Berkeley, before joining the Stanford faculty in 1955. For the last several years, his undergraduate lectures have been televised live to Bay Area industries and colleges.

He has published works in the fields of electromagnetic theory, applied mathematics, radio astronomy, and solar physics, and is interested in the theory of large moving structures—as in the radio telescopes that he built on the Stanford campus.

After writing books on radio astronomy and applied mathematics which have been translated into foreign languages, he himself translated a radio astronomy text from French to English.

With the advent of the space age, Professor Bracewell made observations of Sputnik I, became interested in satellite dynamics, solved the problem of the stability of spinning satellites, and supplied the press with the first accurate predictions of the path of the sputniks over California. Bracewell became one of the early contributors to the post-Sputnik discussions of life in space.

Professor Bracewell belongs to the American Astronomical Society, the Royal Astronomical Society, and is a Fellow of the Institute of Electrical and Electronic Engineers. He was a member of Stanford's Senate and Committee on Committees in 1970-72.